Design Automation and Applications for Emerging Reconfigurable Nanotechnologies

Shubham Rai • Akash Kumar

Design Automation and Applications for Emerging Reconfigurable Nanotechnologies

Springer

Shubham Rai
Department of Computer Science
TU Dresden
Dresden, Germany

Akash Kumar
Department of Computer Science
TU Dresden
Dresden, Germany

ISBN 978-3-031-37923-9 ISBN 978-3-031-37924-6 (eBook)
https://doi.org/10.1007/978-3-031-37924-6

This Springer imprint is published by the registered company Springer Nature Switzerland AG
The registered company address is: Gewerbestrasse 11, 6330 Cham, Switzerland

Paper in this product is recyclable.

Preface

In the last few decades, two major phenomena have revolutionized the electronic industry—the ever-increasing dependence on electronic circuits and the Complementary Metal Oxide Semiconductor (CMOS) downscaling. These two phenomena have been complementing each other in a way that while electronics, in general, have demanded more computations per functional unit, CMOS downscaling has aptly supported such needs. However, while the computational demand is still rising exponentially, CMOS downscaling is reaching its physical limits. Hence, the need to explore viable emerging nanotechnologies is more imperative than ever. This book focuses on streamlining the existing design automation techniques for a class of emerging reconfigurable nanotechnologies. Transistors based on this technology exhibit duality in conduction, i.e., they can be configured dynamically either as a p-type or an n-type device on the application of an external bias. Owing to this dynamic reconfiguration, these transistors are also referred to as Reconfigurable Field-Effect Transistors (RFETs).

Exploring and developing new technologies, just like CMOS, require tackling two main challenges—first, design automation flow has to be modified to enable tailor-made circuit designs; second, possible application opportunities should be explored where such technologies can outsmart the existing CMOS technologies. This book targets the above two objectives for emerging reconfigurable nanotechnologies by proposing approaches for enabling an Electronic Design Automation (EDA) flow for circuits based on RFETs and exploring hardware security as an application that exploits the transistor-level dynamic reconfiguration offered by this technology.

This book explains the bottom-up approach adopted to propose a logic synthesis flow by identifying new logic gates and circuit design paradigms that can particularly exploit the dynamic reconfiguration offered by these novel nanotechnologies. This led to the subsequent need of finding natural Boolean logic abstraction for emerging reconfigurable nanotechnologies as it is shown that the existing abstraction of negative unate logic for CMOS technologies is sub-optimal for RFETs-based circuits. In this direction, it has been shown that duality in Boolean logic is a natural abstraction for this technology and can truly represent the duality in conduction offered by individual transistors. Finding this abstraction paved the way for defining suitable primitives and proposing various algorithms for logic synthesis and technology mapping.

The following step is to explore compatible physical synthesis flow for emerging reconfigurable nanotechnologies. Using silicon nanowire-based RFETs, `.lef`, and `.lib` files have been provided which can provide an end-to-end flow to generate `.GDSII` file for circuits exclusively based on RFETs. Additionally, new approaches have been explored to improve placement and routing for circuits based on reconfigurable nanotechnologies. It has been demonstrated how these approaches led to superior results as compared to the native flow meant for CMOS.

Lastly, the unique property of transistor-level reconfiguration offered by RFETs is utilized to implement efficient Intellectual Property (IP) protection schemes against adversarial attacks. The ability to control the conduction of individual transistors can be argued as one of the impactful features of this technology and suitably fits into the paradigm of security measures. Prior security schemes based on CMOS technology often come with large overheads in terms of area, power, and delay. In contrast, RFETs-based hardware security measures such as logic locking, split manufacturing, etc. proposed in this book, demonstrate affordable security solutions with low overheads.

Overall, this book lays a strong foundation for the two main objectives—design automation and hardware security as an application, to push emerging reconfigurable nanotechnologies for commercial integration. Additionally, contributions done in this book are made available under open-source licenses so as to foster new research directions and collaborations.

Dresden, Germany

Shubham Rai
Akash Kumar

Acknowledgments

A Ph.D. is a transformation, not only does it transform your name with a preceding title but also transforms you as a person. This transformation is very similar to simulated annealing. You go through various recovery and growth phases to become a better individual. With this book, I would like to take this opportunity to sincerely thank everyone who has helped me in various stages of this transformation.

I think the most earnest appreciation is (and will always) be reserved for Prof. Dr. Akash Kumar. First, to accept me as a Ph.D. candidate and then continuously guiding and motivating me throughout this wonderful journey. He has been an exceptional mentor and a great friend. His working style and his encouragement are truly remarkable. He provides complete scientific freedom and yet was also involved in research sometimes even to the level of minute algorithmic details.

I would like to specially mention three people whom I would like to thank from the bottom of my heart—Dr. Jens Trommer from whom I can go and ask about RFETs technology anytime, Dr. Heinz Riener who helped me during my stay at EPFL and from whom I learned a lot about logic synthesis, and last but not the least Dr. Siva Satyendra Sahoo, my badminton buddy and the guy who is always willing to discuss and from whom I have learned all the nitty-gritty details such as writing paper, presenting results, etc. These guys have been instrumental at various stages of my Ph.D.

I would also like to extend my sincere gratitude to the doctoral committee comprising of stalwarts—Prof. Dr. Giovanni De Micheli, Prof. Dr. Thomas Mikolajick, Prof. Dr. Walter M. Weber, and Prof. Dr. Diana Göhringer. I would like to express my appreciation toward my collaborators—Dr. Andreas Krinke, Dr. Andrè Heinzig, Dr. Dennis Walter, and Dr. Satwik Patnaik.

The researchers at the Chair of Processor Design had been an amazing group of people to work with. Martin, with whom I first interacted here in the group, helped me immensely in the beginning when I moved here. Michael with whom I have all my initial RFETs discussions. Conny, for helping me out with all the formalities related to my Ph.D. Tuan and Golsa, for our small coffee group. Salim, with whom I started my Ph.D. journey, our discussions on various topics—ranging from research to family to social life. Ansh, who is just like my younger brother, someone from whom I can ask for any help at any time of the day. My present roommate, Steffen, one of the coolest German I have met. Behnaz, Yuhao, and Zahra with whom I had various kinds of discussions. Thanks you all for making my time invaluable here. Lastly, I would also like to extend my thanks to the graduate and undergraduate students I worked with—Pallab, Aditya, Abhiroop, Martin, Nishant, Yasasvi, and Gopal.

I would also like to thank my family, particularly my parents, who have been a constant support. My mother who was always worried that I don't have the capacity to study. Thanks, mummy for your consistent effort in pushing me throughout my life. My father, who himself holds a Ph.D. degree, has been my first inspiration to pursue this transformation. Bhaiya, bhabhi, Chotu, and my in-laws—thanks for your encouragement at various stages.

The biggest support and the staunchest believer in my abilities has been my wife, without whom I think I would not have even stepped on this journey. She was the one who encouraged me to pursue this and had been always motivating me whenever I was feeling down. She has been very understanding during deadlines when I had to work on weekends and late nights. Special thanks to her for having the greatest gift of our life—our daughter, Avriha.

Shubham Rai

Contents

Acronyms

.GDSII	Graphic Data Stream.
AIG	And-Inverter Graph.
ALU	Arithmetic Logic unit.
BDD	Binary Decision Diagram.
BEOL	Back-End-of-Line.
BTBT	Band-to-Band Tunneling.
CAD	Computer-Aided Design.
CG	Control Gate.
CGRA	Coarse-Grained Reconfigurable Array.
CMOS	Complementary Metal Oxide Semiconductor.
CNF	Conjunctive Normal Form.
CNTFET	Carbon Nanotube Field-Effect Transistor.
DAG	Direct Acyclic Graph.
DFF	D Flip-Flop.
DIGFET	Dual-Independent Gate Field-Effect Transistor.
DNF	Disjunctive Normal Form.
DRC	Design Rule Check.
EDA	Electronic Design Automation.
EOT	Effective Gate Oxide Thickness.
ESOP	Exclusive Sum-of-Products.
FEOL	Front-End-of-Line.
FET	Field-Effect Transistor.
FPGA	Field-Programmable Gate Array.
GAA	Gate-All Around.
GeNW	Germanium Nanowire.
HD	Hamming Distance.
HOF	Higher-Order Function.
HPWL	Half-Perimeter Wire Length.
HVT	High Vt (Threshold Voltage).
IC	Integrated Circuits.
IoT	Internet-of-Things.

IP	Intellectual Property.
LUT	Look-Up Table.
LVT	Low Vt (Threshold Voltage).
MIG	Majority-Inverter Graph.
MIGFET	Multi-independent Gate Field-Effect Transistor.
NFET	n-Channel Field-Effect Transistor.
NMOS	n-Channel Metal Oxide Semiconductor.
OER	Output Error Rate.
P&R	Placement and Route.
PDK	Process Design Kit.
PDN	Pull-Down Network.
PFET	p-Channel Field-Effect Transistor.
PG	Program Gate.
PMOS	p-Channel Metal Oxide Semiconductor.
POS	Product-of-Sums.
PSO	Power Shutoff.
PUF	Physically Unclonable Function.
PUN	Pull-Up Network.
QoS	Quality of Service.
RFET	Reconfigurable Field-Effect Transistor.
SB	Schottky Barrier.
SBFET	Schottky Barrier Field-Effect Transistor.
SiNW	Silicon Nanowire.
SOC	Silicon-On-Chip.
SOI	Silicon-On-Insulator.
SOP	Sum-of-Products.
TCAD	Technology Computer-Aided Design.
TFET	Tunneling Field-Effect Transistor.
TIGFET	Three-independent gate Field-Effect Transistor.
TLL	Transistor-Level Locking.
TMD	Transition Metal Dicalchogenide.
TSPC	True-Single Phase Clock.
XAG	Xor-Inverter Graph.
XMG	Xor-Majority Graph.

List of Figures

List of Tables

Chapter 1
Introduction

At the time, when I started writing this book, the iPhone 13 series was being launched. And the most striking thing, for me as a researcher, is that the new-generation iPhones have been able to shrink the CMOS channel width by yet another 2 nm. They used a 5-nm technology node which allowed them to use 27% more transistors than the previous generation. Still, this progression is not too far from the quintessential Moore's law prediction [Moo+65]. Although area scaling continues roughly at the rate of 0.5 times every two years [BY17], the cost to build transistors and the power consumption per unit transistor have not been able to follow the promises of Moore's law and Dennard scaling methodology [Den+74]. This trend has not been abrupt and Moore's law that has been guiding and predicting transistor downsizing from the last five decades is really pushing the physical limits of the transistor's channel width, as 1 nm is equivalent to the thickness of just 5 silicon atoms!

Transistor downscaling refers to the decrease in the size of the channel length of individual transistors. This has a clear impact on the resistance and power consumption by individual transistors. Particularly at lower-technology nodes, power consumption and heat management issues in high-performance integrated circuits are huge problems. It is not about the crude reality of transistor downscaling, but it is about the real physical challenges that lie in the transistor's geometry and routing of metal interconnects at low-technology nodes. In spite of these issues, two major factors that allowed CMOS to flourish in the past and the present are its increased circuit robustness and reduced power dissipation with every new generation.

While the exploration to reach the physical limits of 1 silicon atom width for integrated circuits still continues, another plausible approach is to look for emerging technologies that can bridge the requirements of future electronic circuits with their unique feature sets [Rai+18b]. The efforts are not directly aimed at replacing CMOS technology for future electronic circuits but to investigate the possibility of using emerging technologies to complement (or supplement) the

S. Rai, A. Kumar, *Design Automation and Applications for Emerging Reconfigurable Nanotechnologies*, https://doi.org/10.1007/978-3-031-37924-6_1

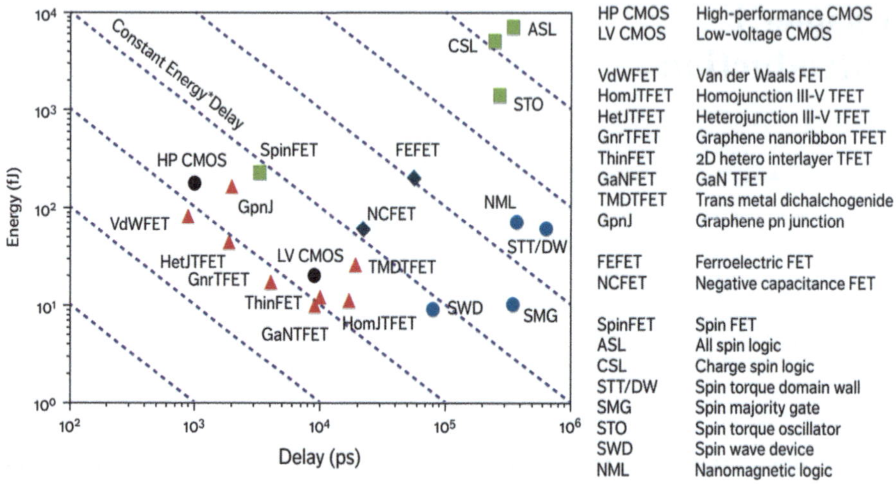

Fig. 1.1 Various emerging technologies have been compared in [BY17]. The figure shows switching energy and delay for a 32-bit arithmetic logic unit circuit

CMOS technology. Figure 1.1 shows various emerging technologies compared with both *high-performance* and *low-power* CMOS technology. All these emerging technologies are compared in terms of switching energy and delay. The figure shows a clear perspective on various technologies and indicates which technologies are worth exploring in terms of these two parameters. While this figure showcases experimental data for various technology, another crucial factor that drives the adoption of a specific emerging technology is the ease of fabrication and coherence with the existing fabrication setup of CMOS. Establishing a new foundry setup requires a multibillion-dollar investment. Hence, an emerging technology can be readily adopted if there is a partial or full overlap with the established setup in terms of their manufacturing. And this is where the *emerging reconfigurable nanotechnology* can be a game-changer as it promises new computational capabilities and is also compatible with the existing top-down fabrication stack of the contemporary CMOS [Sim+16, Mik+17].

One of the decisive prerequisites to efficiently integrate emerging nanotechnologies into commercial electronics is the consideration of the physical properties of the technologies within the EDA flow [Ama+15b, Rai+18b]. Devising efficient flows is necessary to accelerate the commercial feasibility of newer nanotechnologies. The recent development of a modern RISC-V processor made with Carbon Nanotube Field-Effect Transistors (CNTFETs) is one such example of a practical circuit based on *beyond-CMOS* technologies [Hil+19]. The constant debate over emerging nanotechnologies compared to existing silicon-based CMOS technologies and the complexities involved in the overall design flow has so far precluded realization and development of strong examples for electronic circuits. The work done in [Hil+19] stands out as a stepping stone to solve the ever-increasing problem with CMOS dimension scaling and the growing skew between cost and performance for CMOS-based circuits [BY17]. Hence, exploring emerging nano-devices is not just an academic exercise but an imperative demand to meet the requirements of future electronics [Rai+18b].

In this direction, the book explores the feasibility of emerging nanotechnology called *reconfigurable nanotechnology* through the lens of design automation. Reconfigurable nanotechnology boasts of a remarkable feature where individual transistors can be tuned to either demonstrate a p-type or an n-type functionality. Within the research community, such devices are termed as Reconfigurable Field-Effect Transistor (RFET) or polarity-control devices. Apart from exhibiting transistor-level dynamic reconfiguration, these devices offer flexibility in terms of power and come with low leakage power dissipation [Mik+17]. Efficient circuit designs have been proposed in the literature exploiting these exciting properties [Rai+17, Tro+16, Gai+13]. However, the nonexistence of a well-formalized EDA flow to build circuits with RFETs opens up a major research direction which is still under-explored. Owing to these special benefits offered by reconfigurable nanotechnology, this book devises new techniques and approaches for enabling design automation for RFET-based circuits such as defining logical abstraction for logic synthesis flows, designing standard cells for technology-independent mapping, and formalizing a physical synthesis flow compatible with existing industrial tool flows for RFETs-based circuits. The circuit design paradigms drive efforts in EDA to propose RFET-centric synthesis flow that (i) gives trustworthy predictions for circuits based on RFETs and (ii) provides complete design flow which can be used to build actual circuits based on RFETs. The book also investigates suitable application scenarios for such an exciting technology and maps the technology's feature set with the demands of *hardware security*. It proposes various circuit design paradigms and lays down approaches that can be used in hardware security applications.

1.1 What Are Emerging Reconfigurable Nanotechnologies?

Emerging reconfigurable nanotechnology is driven by transistors that exhibit *ambipolarity* at the transistor level. Ambipolarity is a property by which both charge carriers—electrons and holes—can tunnel through an electrostatic barrier. With process techniques, ambipolarity can be enhanced to demonstrate electrical symmetry in both types of conduction, thereby enabling transistors to demonstrate both p- and n-type properties independently on the application of external bias potential. This enables dynamic reconfiguration of individual transistors either as a p-channel Field-Effect Transistor (PFET) or n-channel Field-Effect Transistor (NFET). This dynamic reconfiguration in RFETs is not facilitated using chemical doping caused by impurities but rather due to electrostatic doping [Mik+17] caused by an external bias potential. Exemplary I-V characteristics for an RFET is shown in Fig. 1.2 [Hei+13]. The right figure shows near-symmetrical I–V characteristics for both p- and n-type configurations. Reconfigurable transistors

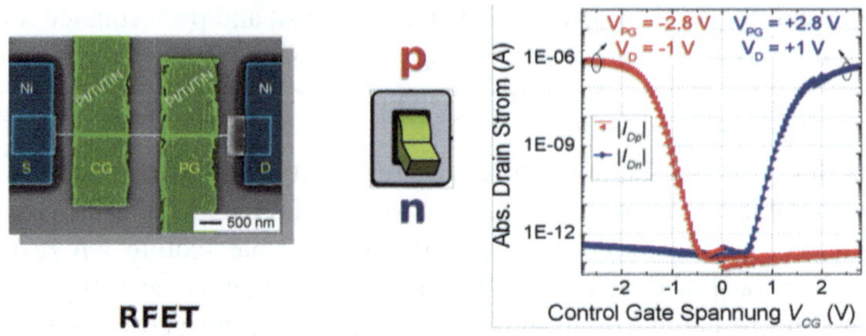

Fig. 1.2 Current-voltage characteristics showing reconfigurable device properties. The switch analogy demonstrate the dynamic reconfiguration between p- and n-type behavior. Experimental data showcasing near-symmetrical electrical properties for both p- and n-type behavior

encapsulate the functionality of two different devices (PFET and NFET) into a single device. The symmetric I–V characteristics allows (re-)programmability of transistor's conductivity as a device that *switches* its properties from p-type to n-type functionality [Tro+16].

The special property of dynamic reconfiguration enabled by metal-semiconductor-metal contacts further allows symmetric electrical conduction in both p- and n-type behavior [Tro+15]. RFETs, just like CMOS devices, come with the same set of terminals—source, drain, bulk, and gate. However, unlike CMOS, RFETs have two types of gate terminals—the *program gate* (PG) and the *control gate* (CG). The Control Gate (CG) is the same as that in the case of CMOS that controls the flow of charge carriers, while the PG controls the type of the charge carriers flowing through the channel. The bias applied at the PG configures the transistor to function either as a p-type or n-type device. Additionally, transistors belonging to reconfigurable nanotechnology allow multiple gate terminals on a single channel [Sim+18]. The possibility of having multiple gate terminals on a single channel within a reconfigurable device opens up new circuit design opportunities that were not possible using the classical Field-Effect Transistor (FET). The CMOS-style complementary pull-up and pull-down networks are, thus, enriched by the abovementioned features and can be tuned to provide more functionalities per computation unit. To address the atavistic excitement of a reader, a simple manifestation of such interesting properties in a circuit design is to have NANDs and NORs with equal performance and equal area.

1.2 Why Does This Technology Look So Promising?

The combination of dynamic reconfiguration between p- and n-type behavior at the transistor level and the possibility of having multiple gate terminals on a single channel open up new design opportunities for circuits based on reconfigurable nanotechnology. First, since the p- and n-type configurations exhibit symmetrical electrical conduction, pull-up and pull-down networks offer equal current drive strengths. Hence, unlike CMOS technology, effort for transistor sizing can be saved in the case of circuits based on RFETs. This symmetry in conduction for both p- and n-type configuration allows flexibility during (i) the design of circuit topologies [Rai+17, GAM14], (ii) the design of layouts for standard cells [BM15, Reu+21], and (ii) the placement and routing of individual cells [Kri+21]. Second, having multiple gate terminals on a single channel reduces the channel on resistance per input [Tro+16]. This allows multiple inputs to be connected to a single transistor, thereby allowing efficient logic gate designs with more than two inputs. Additionally, multiple gate terminals also play an important role in controlling the performance of a logic gate [Zha+14].

With the freedom to choose transistor's electrical behavior at runtime and the possibility of designing logic gates with more than two inputs, contemporary CMOS-styled circuit design is not optimal for RFETs and requires rethinking in terms of logic gates and circuit designs [Mik+21]. The flexibility to alter the properties of an individual transistor either to add more inputs or to increase the performance finds extensive application in electronic circuits [Rai+18b]. This dynamic reconfiguration of electrical properties allows circuits or logic gates to exhibit multiple yet mutually exclusive functionality with runtime reconfiguration that finds huge applications in both contemporary and future applications such as hardware security [RRK18, Rai+20b, Bi+17] and reconfigurable computing fabric [DW04, Gai+15].

1.3 Electronic Design Automation

EDA comprises of various tools, algorithms, and approaches required to build an electronic circuit from a given specification of an application. An EDA flow starts with a description or a specification of an application and uses the above tools to realize an actual physical layout.

A typical EDA flow is shown in Fig. 1.3. It primarily consists of two main synthesis stages—*logic synthesis* and *physical synthesis*. Logic synthesis maps a logic graph representing a circuit description onto some logic primitives. These primitives can be either standard cells that represent logic gates in a typical *application-specific integrated circuit* (ASIC) flow or look-up tables (LUTs) in an FPGA flow. Logic synthesis uses various algorithms and heuristics to reduce the size (number of nodes in the logic representation) and depth (number of levels between the primary input and primary output) of the starting logic network. Logic synthesis, thus, has a huge impact on the final area and the delay of the circuit. Logic synthesis primarily deals with logic minimization and optimization stages as shown in Fig. 1.3. In this book, we focus only on the standard-cell-based ASIC flow.

After obtaining an optimized logic network, technology-independent mapping generates a netlist using a library of standard cells. The technology mapping stage uses various heuristics [CWD99, MBV06] to map the netlist of a circuit onto some

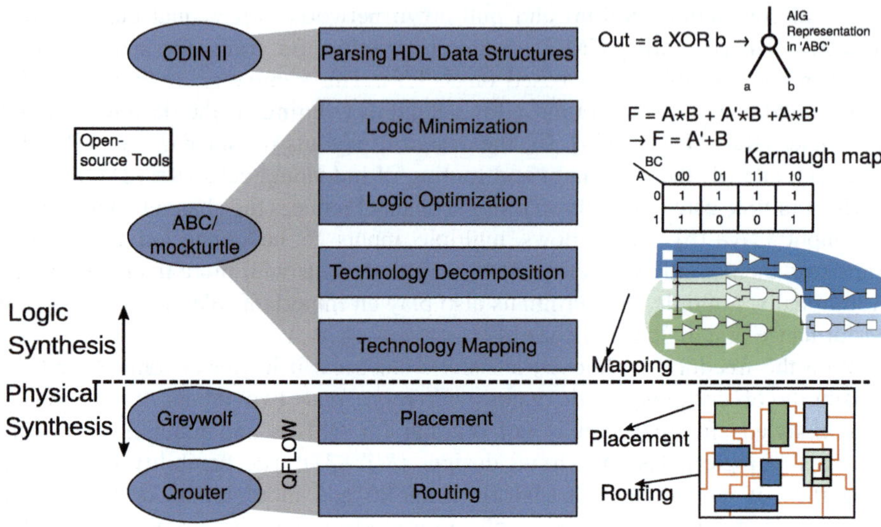

Fig. 1.3 Various stages in EDA

logic primitives so as to minimize area, delay, or both. Physical synthesis uses this technology-mapped netlist and converts it into a final physical layout that is fabrication-ready. It translates a network of logic cells into a chip layout with actual physical standard cells based on a particular underlying technology to be handed over to the foundry for fabrication. As shown in Fig. 1.3, physical synthesis flow consists of two important steps of *placement* and *routing* (P&R). Placement carries out a *floor planning* of the available *die* area and places the connected logic cells of the technology-mapped netlist onto the die area. It considers various metrics so that the placement of logic cells is such that the connection between two parts of the circuit is not hampered by long metal wires. Routing further aims to minimize the actual wirelengths to reduce wire delays. After the placement and routing, the physical synthesis stage generates a Graphic Data Stream (.GDSII) file to be given to the foundry.

Both the logic and physical synthesis consist of multiple substages that are interlinked to each other. For the sake of automation of these steps, these substages are somewhat abstracted in a way to deal with a single set of problems. For example, logic synthesis deals with logic optimization and minimization and does not consider placing different logic gates together to reduce the wirelength. Similarly, placement deals with the optimal placement of logic cell components on a die as bad placement often leads to high parasitics due to connected cells as well as long wirelengths. Throughout this flow, Boolean equivalence is necessary so that the optimized network is equivalent to the original starting network.

Most of the problems in the EDA domain are computationally intractable, i.e., it is difficult to reach an optimal solution with polynomial-time complexity [Mic94]. Hence, most problems in EDA are solved through heuristics that perform well over certain subclasses of problem. Hence, multiple heuristics have been proposed over the last three decades that have pushed EDA tools for more efficient circuit implementations based on CMOS. Recently, various learning approaches are being explored that can cater to a particular goal or a cost function to further empower the EDA tools [Pan18, Bud+22, Per+21, Rai+21b].

From the last 3–4 decades, EDA has been playing a phenomenal role in utilizing the benefits of every new generation of CMOS technology and translating them into actual circuits. In fact, the very existence of Moore's law has been enabled by advancements in EDA. The quality of the final physical circuit is governed by how well the approaches and algorithms within EDA have been able to optimize the given hardware description. Now with CMOS downscaling and the emergence of new nanotechnology, an important challenge that stands out for EDA tools is to extract even better performance and quality of circuits than earlier [Ama+15b].

One of the biggest impending issues to enable support of emerging nanotechnologies is that the contemporary EDA tools are based on CMOS logic primitives, i.e., most of the algorithms or heuristics are tuned to develop optimal circuits primarily based on CMOS. However, with the emergence of competing transistor technology, the whole stack of EDA faces a tough challenge to incorporate the enhanced functionality from these technologies within the EDA flow. There are several facets involved here. First, modeling of the functionality offered by these technologies is required so that such models can be utilized by various algorithms in EDA. For example, in order to incorporate a power gating approach for particular parts of the circuit, placement algorithms need to handle different power domains in their placement algorithm to enable/disable power gating in those parts of the circuit. Second, new logic synthesis primitives have to be developed which can utilize the benefits of emerging technologies at the logic representation level [Ama+15b]. Third, new design and layouts of standard cells are required for logic gates based on emerging technology [Kri+21].

In the next section, we look at various challenges that need to be addressed to push emerging reconfigurable nanotechnologies for commercial adoption.

1.4 The Game of Seesaw: Key Challenges vs Benefits for Emerging Reconfigurable Nanotechnologies

Any emerging technology faces multiple challenges during its development cycle. For a commercial adoption of a given emerging nanotechnology, challenges and benefits occupy the two sides of a seesaw. From a technology point of view, benefits come to attention first because as an emerging technology, that is where they stand apart from conventional technology. Once the given set of benefits is recognized that a given technology can suitably complement the requirements of a given application, research efforts are required to address the associated challenges. With emerging reconfigurable nanotechnology, multiple challenges restrict their adoption in mainstream electronics.

In terms of predicting whether a particular technology will be commercially viable, one of the seminal paper [Bor06] that looked at electronics from a futuristic point of view laid down three tenets that are essential for any emerging technology to survive and exist. These three tenets are as follows:

1. Gain: All major transistor-based technology showcased either current or voltage gain. This is important to realize a logic fan-out. A single logic gate must be able to drive other logic to constitute a circuit. Technologies going back from vacuum tubes to bipolar transistors to CMOS all exhibited some kind of current or voltage gain.
2. Signal to noise: The signal gain through a logic based on any underlying technology should be more than the ambient noise. This is important so as to avoid auxiliary techniques to extract the useful signal.
3. Scalability: Scalability in terms of physical parameters (such as area, power, or delay) is essential as it allows you to pack more functionality, or provide higher performance or lower energy with every new generation. The perfect example for this is CMOS downscaling, which provides all such benefits and is a major reason that CMOS is still going strong even after 30 years!

RFETs share a lot of properties with contemporary CMOS technology. First, just like CMOS, RFETs demonstrate voltage gain. This voltage gain has been demonstrated through Technology Computer-Aided Design (TCAD) simulation. Second, for the case of signal-to-noise ratio for RFETs, a similar argument can hold as the signal gain is more than the ambient noise, albeit with different observable gains for both p- and n-type configurations. Third, RFETs will also provide similar scalability with every generation due to their overlap with CMOS technology in terms of the top-down manufacturing process.

However, one of the biggest challenges faced by RFETs is the asymmetrical behavior of the VG-IDS characteristic which is typical for all Schottky-barrier Field-Effect Transistors (SBFETs) with metal-semiconductor junctions [Mik+17]. This can be ascertained due to different tunneling probabilities for electrons and holes. However, various approaches ranging from tuning the tunneling probabilities to work function adjustments can be used to achieve acceptable symmetries. Scaling trends with RFETs also hold few concerns, primarily due to two main factors: (i) how downscaling impacts the transistor designs for RFETs considering the transistors need to deliver near electrical symmetry and (ii) how to integrate multiple gate terminals on the channel for the future generation.

While the above tenets are more or less abstracted from the electrical properties of individual transistors, techniques for circuit integration to enable widescale implementation are a major milestone to achieve. Next, we look at these challenges associated with emerging reconfigurable nanotechnology in detail.

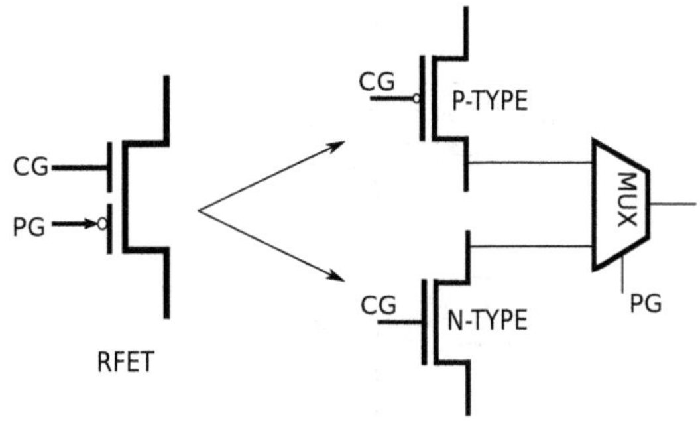

Fig. 1.4 RFET-based logic gate and comparison with CMOS

1.4.1 Abstracting Ambipolarity in Logic Gate Designs

Ambipolarity is the main physical reason which is responsible for transistors to exhibit dynamic reconfiguration between p- and n-type behavior. This can be very well represented as shown in Fig. 1.4. Functionally, an RFET encapsulates the two kinds of MOSFETs, followed by a multiplexer. The select line of that multiplexer is the PG terminal of the RFET. Hence, an RFET can be logically configured either as an NFET or a PFET by steering the bias at the PG.

This transistor-level reconfigurability transcends into logic gates and circuit designs. The paradigm of using complementary networks also works in the case of RFETs. Complementary networks in a typical static CMOS-based logic gate design is characterized by separate pull-up and pull-down networks that are responsible to realize logic 1 and logic 0, respectively. This can be seen in Fig. 1.5a. On the same lines, individual RFETs can be configured in such a way that a separate pull-up and pull-down network can be realized and similar CMOS-styled logic gates can be designed. However, conventional CMOS circuit designs when applied directly with this emerging nanotechnology often result in suboptimal designs. This is attributed to the fact that individual RFET transistors are both larger and slower as compared to CMOS transistors. Hence, such logic gate designs are not practical since it raises the question: why opt for RFETs when similar logic functions can be realized with faster and smaller CMOS transistors?

Hence, new approaches delivering tailored circuit designs are needed to truly tap the exciting feature set of these reconfigurable nanotechnologies. Logic gate designs that are able to exploit the inherent ambipolarity can offer better performance and area as compared to CMOS-styled designs. Using transistor-level reconfiguration, RFETs-based circuit designs can use the switching capabilities of RFETs to interchange pull-up and pull-down networks simultaneously. The interchangeable pull-up and pull-down network as shown in Fig. 1.5b enabled by ambipolar RFETs allow efficient circuit designs based on RFETs [Rai+20a].

Such dynamic reconfiguration is not new in electronics, and an *arithmetic logic unit (ALU)* is a perfect example of such systems where a user has an option to choose from a set of arithmetic functions. The whole ALU function can be seen

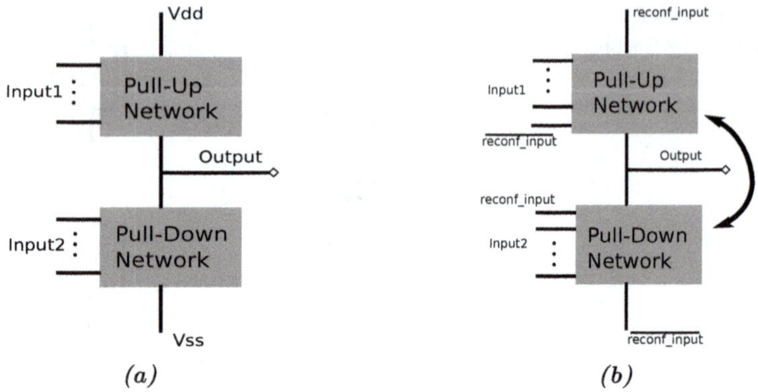

Fig. 1.5 (**a**) Fixed pull-up and pull-down network in the case of complimentary MOS logic gates. (**b**) Interchangeable pull-up and pull-down network in case of RFET-based logic gates. The reconf_input decides the logic functionality

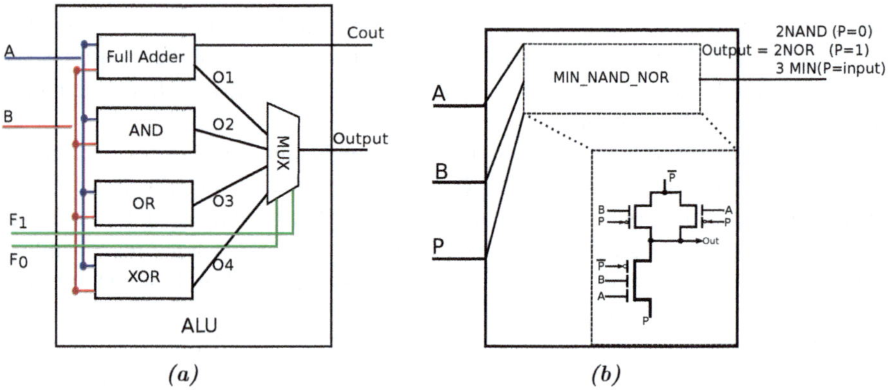

Fig. 1.6 Reconfigurability in terms of logical abstraction. (**a**) Extrinsic reconfigurability. (**b**) Intrinsic reconfigurability

as a function f encapsulating functions like *addition* and *multiplication* as g and h, respectively. This kind of reconfiguration is available because of extra circuitry and such circuits are characterized with multiple control paths that can give more than one function simultaneously. Such reconfigurability can be termed as *extrinsic reconfigurability*. This is shown in Fig. 1.6a. We can see that each inbuilt logic function does produce multiple outputs simultaneously through multiple control paths shown as O1, O2, O3, and O4. In the end, the *MUX* is used to select the required output. It can be seen from the figure that logic functions from each logic gate are implemented independently and their selection is done by MUX at runtime.

The enhanced functionality exhibited by RFET differs from that of a configurable circuit like ALU as there is a *mutual exclusion* in the availability of multiple functions for RFETs. By mutual exclusion, we mean that an RFET-based logic gate can have one and only one logical output at a single instant of time, unlike extrinsic reconfigurability. This mutual exclusion is the result of the way input variables are connected to the gate terminals of the *HOF*. This is shown in Fig. 1.6b. Unlike extrinsic reconfigurability, there is neither extra circuitry involved here nor are there multiple output paths followed by a selection logic. The unique behavior is possible due to the electrical properties of RFETs. Such kind of reconfiguration can be termed as *intrinsic reconfigurability*.

Intrinsic reconfigurability can be used to implement circuits with more than one functionality. Reconfigurability can thus be used as a radical measure to tackle CMOS downscaling by realizing more functionality per computational unit. However, reconfigurability often comes with delay and area overhead. Hence, circuit design techniques should be aware of such overheads.

1.4.2 Enabling Electronic Design Automation for RFETs

EDA plays one of the most important roles in enabling the commercial integration of an emerging technology. It provides a bridge from laboratory-level explorations of an emerging technology to an actual physical realization in the form of a circuit.

Reconfigurable nanotechnology boasts many flexibility-friendly properties, thereby altering the size, performance, and power consumption of circuits based on reconfigurable nanotechnologies. They hold great promise as suitable primitives for enabling multiple functionalities per computational unit. Handcrafted RFET-based logic gate designs showcased huge improvements in terms of area and delay over the contemporary CMOS-style designs [Rai+17, Gai+13]. The major limitation, however, is to bring such novelty in circuit designs within the fold of design automation techniques. From a circuit integration point of view, there are a few challenges—first, an individual RFET transistor is bigger as compared to a CMOS transistor at the same technology node (because of the extra gate terminal). To compensate for the area overhead, circuit design approaches should be explored which can utilize the high functional expression offered by RFETs. Second, which design topologies can exploit the dynamic reconfiguration between pull-up and pull-down networks? In this direction, is there any logical abstraction that can guide circuit design exploration for RFET-based circuits? Third, which application can particularly use or exploit the properties of RFETs to drive the overall development of RFETs as a potential technology? The various tools and approaches within EDA should interact and reciprocate the properties of reconfigurable nanotechnology. RFET's unique properties need to be abstracted in a way such that it can be handled during the synthesis stages as described earlier in Sect. 1.3.

An early evaluation in terms of circuit design is also essential to assess the feasibility and practicability aspects of emerging nanotechnologies. This can be only done through EDA tool flows over a benchmark suite to have acceptable estimates. This also helps to provide feedback to the technology designers to gauge flaws in the transistor designs. This early evaluation has to be done both at the physical synthesis and logic synthesis levels. For example, early handcrafted designs of logic gates demonstrate that RFETs can realize switchable pull-up and pull-down networks that can allow reconfiguration as shown in Fig. 1.5b. This hugely impacts designing logic gates based on RFETs and also is a defining factor as to how technology mapping needs to be modified to abstract the switching potential of the pull-up and pull-down networks. Defining logical abstraction to encapsulate the inherent ambipolarity is necessary to push for further gains [Ama+15b].

Most of the attempts in enabling EDA for RFETs focused to model manual designs of logic gates and simple circuits into the fold of either logic synthesis or technology mapping [AGM13, Ama+15a, AGD14]. Apart from abstracting ambipolarity, EDA (particularly physical synthesis) has to tackle new RFET-based layouts. This requires designing new physical synthesis flows which can compensate the area overheads since an individual RFET is larger as compared to CMOS. RFETs also come with an additional gate terminal which can lead to routing congestion. Hence, delay overheads due to excess routing resources need to be evaluated.

Scalability, which has been portrayed as one of the tenets earlier, is realized to a great extent by EDA techniques. Design automation techniques should be able to tap the benefits of a single device and map them to the application requirements. Techniques to explore design automation for circuits based on RFETs are required to investigate its commercial feasibility. A strong interaction between the above challenges and EDA approaches can catapult this technology for commercial adoption.

1.4.3 Enhanced Functionality: A Suitable Fit for Hardware Security Applications

With the rise of neuromorphic computing and Internet-of-Things (IoT) applications, there is a growing trend of specialization of hardware for specific tasks [Sha20]. The demand for specialized hardware to tackle a specialized application requirement has opened up new avenues in the search for emerging nanotechnologies. Several emerging technologies such as ReRAMs, spin-based devices, and FeFETs find huge applications in crossbar architecture, nonvolatile memory, and neuromorphic architectures, respectively [Rad+, Rai+21a, YC16]. An emerging technology that promises to solve one of the biggest limitations of a crucial application can be readily adopted. The prime reason driving this trend is that the current hardware architecture is limited from the physical point of view. Let us take the example of neuromorphic computing which needs a huge amount of both memory and computational resources. Conventional Von Neumann architecture is characterized by separate memory and processing elements where the transfer of information can be a bottleneck in increasing throughput. In this case, the property of FeFETs to allow simple mathematical operations in the memory itself plays a major role in reducing

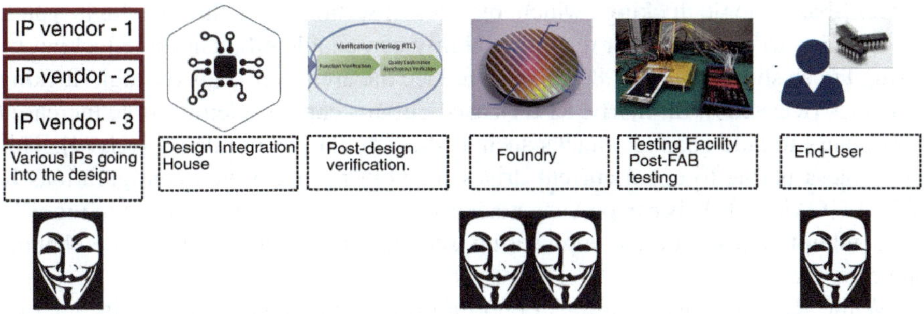

Fig. 1.7 Global IC supply chain and hardware security attacks possible at various phases

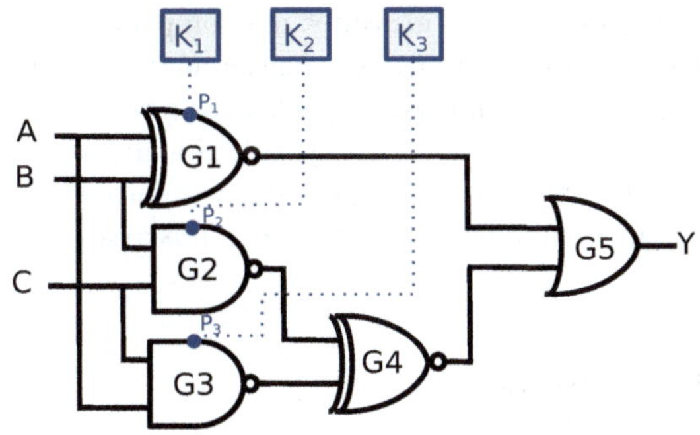

Fig. 1.8 Logic locking using RFETs, where the PG acts as the key input. Realizing logic locking in conventional CMOS necessitates insertion of additional logic gates whereas the inherent construction of RFETs facilitates non-insertion of additional logic gates. The PG signals are driven from an on-chip *tamper-proof* memory

the communication between memory and the processing element, thereby greatly improving the computation paradigm for neuromorphic computing [Bas+21].

Hardware security has gained importance in the last decade due to widespread globalization of the Integrated Circuits (IC) supply chain. Hence, ICs are susceptible to different attacks at various phases of the global supply chain as shown in Fig. 1.7. Ensuring security in ICs is an unavoidable cost not only in terms of capital required but also at the circuit level, as it often comes with associated overheads.

Owing to the dynamic reconfiguration at the device level, RFETs open up new avenues for efficient yet cost-effective solutions to implement hardware security features for circuits [Kne20]. The freedom to configure a transistor either as a p-type or an n-type, depending upon the requirement, allows constructing polymorphic logic gates at lesser overheads in terms of area, power, and delay than the conventional CMOS. The dynamic reconfiguration allows the circuit to be reconfigured at runtime to deliver a particular functionality from a range of given functions. Such polymorphism can suitably fit within the paradigms of logic locking schemes [Raj+15, Mas+17, Cha+20] as shown in Fig. 1.8 [Rai+20b]. Here, unlike

CMOS-based logic locking, which requires insertion of additional logic gates, RFETs due to their inherent reconfigurability can enable efficient logic locking by using PG as shown in Fig. 1.8. This can reduce the area overheads for logic-locking schemes [Rai+20b]. Similarly, RFET-based circuits can also demonstrate increased robustness to side-channel attacks such as differential power attacks [KJJ99]. This robustness is due to equal current drives for both p- and n-type configurations in RFETs [GMT22]. This can prevent attackers at the end user level (shown in Fig. 1.7) who intend to use reverse-engineering schemes involving various side-channel attacks.

While RFETs provide various opportunities for implementing security guarantees in a given electronic circuit, designing circuits utilizing these properties remains a major challenge. Analysis of circuits in terms of implemented measures is the next step to ensure whether the security measures are comparable to the state of the art or not. Lastly, quantitative predictions and measurements should be carried out to give estimates in terms of area, power, and delay for a conventional CMOS circuit to motivate the adoption of newer technology. Additionally, since reconfiguration is central to RFET-based circuits, is there a possibility of *misconfiguration* in RFETs? If yes, what are the electrical repercussions of such misconfiguration in a circuit?

1.5 Research Questions

With CMOS device scaling reaching the physical and economic limits, emerging reconfigurable nanotechnologies offer multiple flexibility-friendly options for circuit designs. Based on the above discussions and challenges, this book aims to address the following research objective:

> Explore and devise approaches to complement contemporary EDA tools for designing efficient yet secure circuits based on emerging reconfigurable nanotechnology

As the contemporary EDA tools are catered specially for CMOS technology, devising efficient EDA flows for RFET-based circuit require looking at transistor-level reconfigurability and abstracting it at every stage of a typical design flow as shown in Fig. 1.3. Hence, the above research objective can be broken down into the following research questions:

1. What kind of design topologies should be employed for logic gates and circuits based on RFETs? Can ambipolarity be abstracted for sequential or metastable circuits?
2. Logic gates based on CMOS are governed by electrical characteristics of distinct pull-up and pull-down networks. Is the same CMOS-styled circuit design applicable for RFET-based logic gates? Why can only certain logic functions be implemented with RFETs as functionality-enhanced logic gates? Is there a universal Boolean property that can be implemented efficiently using RFETs?

3. As CMOS favors negative unate Boolean logic, AIGs are the natural abstraction for CMOS logic [Ama+15b]. However, are AIGs appropriate for RFET-based circuits as well?
4. Considering RFETs to be still in a nascent stage in terms of their maturity compared to CMOS, how feasible is it to make actual physical circuits based on RFETs? What are the research and engineering efforts required to formalize a physical synthesis flow for an RFET-based circuit?
5. How polymorphic logic gates can be designed by utilizing transistor-level reconfigurability offered by RFETs to develop efficient and secure circuits? Is the security offered by RFETs strong enough to tackle the state-of-the-art security techniques? And is security the only gain, or do the RFETs-based circuits have certain vulnerabilities?

These research questions are directed towards the main research objective as mentioned before. If we look closely, the research questions are aimed at investigating whether the CMOS-centric EDA approaches are "good enough" for RFETs. At the prima facie, the CMOS-style flows and CMOS-style logic gates do provide a good starting point for RFETs. However, the subsequent chapters demonstrate that (i) RFET-centric flows can give remarkable improvements as compared to the default CMOS-style flows and (ii) development of RFET-centric flows is done in cognition with the contemporary CMOS flows so that it can be readily adopted. This is also done to reduce the development cost by focusing on the modeling of RFET's specific properties within the EDA flow. The work done in the book is aimed at providing synthesis flows available under an open-source license that can benefit the research community.

1.6 Entire RFET-Centric EDA Flow

Figure 1.9 shows the overall EDA flow for RFETs-based circuits aimed at tackling the research objective mentioned in the previous section. The blue arrows are characteristic of a typical EDA flow. It shows how, through various stages in an EDA flow, a hardware description is converted to the final physical layout based on RFETs. The green arrows show how the reconfigurable property of RFETs can be integrated or abstracted within the EDA flow. The close interaction between the two flows describes the contribution of this book.

This close interaction starts at the lowest abstraction with transistor models for RFETs. There are multiple device candidates demonstrating reconfigurable properties. Indeed, various RFET models come with structural nuances that make them distinct from each other. These models are introduced in Chap. 2. From a logical perspective, they all demonstrate ambipolarity that can be manifested as a transistor-level reconfiguration. Using the transistor-level reconfiguration, functionality-enhanced logic gates are presented in Chap. 3. Efficient multifunctional combinational and sequential logic gates and circuit implementations are

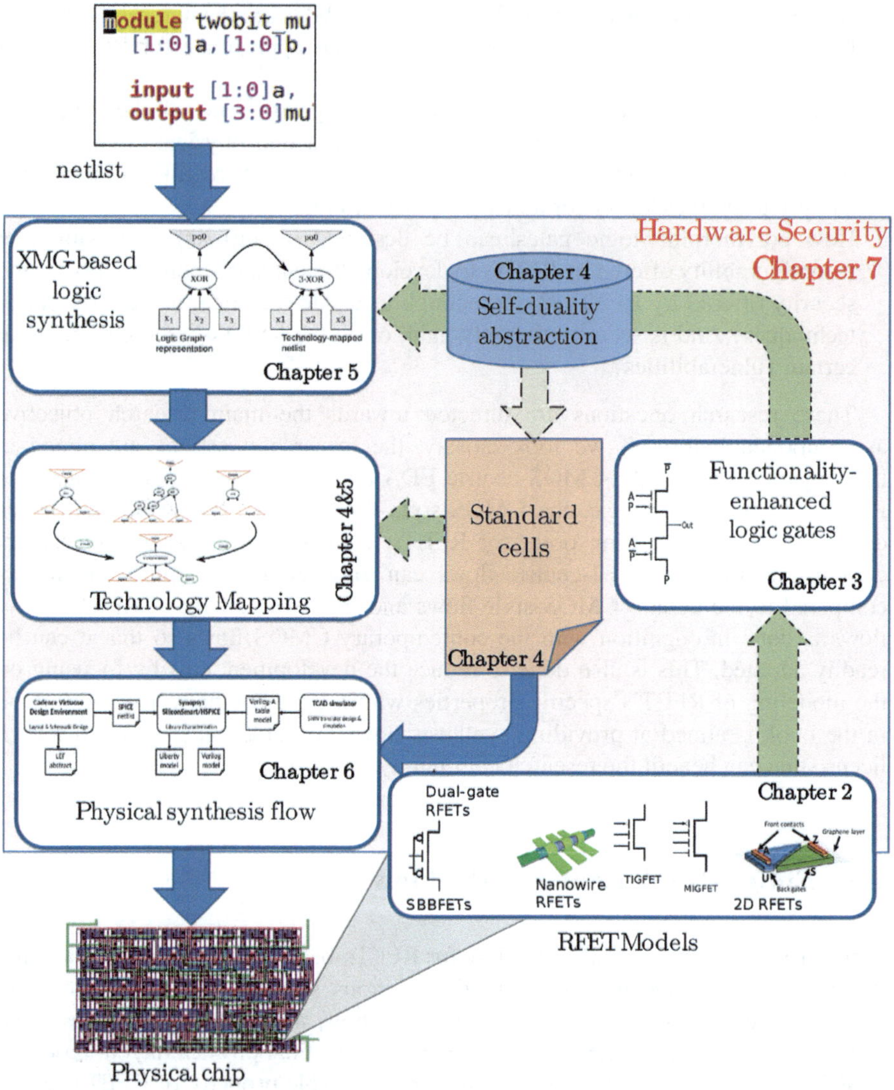

Fig. 1.9 Entire RFET-centric EDA flow starting from a given netlist to the final layout of the circuit

proposed. Particularly in sequential logic gates, ambipolarity can play a huge role in increasing throughput for sequential circuits. Calculation of normalized area and delay for these logic gates demonstrate better performance and reduced area as compared to the CMOS technology. These functionality-enhanced logic gates enable new design approaches for RFET-based circuits.

These logic gates form the basic foundation to explore logical abstraction for RFETs as it raises the question of what kind of logic functionality can gain from the ambipolar device characteristics. These functionality-enhanced logic gates share a specific Boolean property where they all demonstrate better logic realization than the contemporary CMOS design in terms of the number of transistors and performance. The interchangeable pull-up and pull-down network as shown in Fig. 1.5b enabled by ambipolar RFETs allow efficient circuit designs based on RFETs. This *interchangeability* is possible only with logic functionalities having the Boolean property of *self-duality*. Thus, self-dual Boolean logic is recognized as the natural abstraction for RFETs and logic functionalities. The formalization of Boolean self-dual property as the logical abstraction for RFETs is covered in Chap. 4. With self-duality as the basic Boolean property, an algorithmic approach has been proposed to distill all logic gates up to five inputs. These logic gates can be utilized to make RFET-based standard cells for circuit designs. Additionally, since RFET-based logic gates can demonstrate more than one functionality, the concept of Higher-Order Function (HOF) is proposed to the abstract functionality of logic gates. This has been embedded in the ABC technology mapper [BM10] to carry out an early evaluation over a benchmark suite to give estimates in terms of area and delay overheads using functionality-enhanced logic gates.

With the Boolean property of self-duality established as the natural logical abstraction, a direct correlation can be established between the available self-duality of the circuit and the area occupied after technology mapping. This is investigated in Chap. 5 where it is shown that better area reduction for RFET-based circuits is achieved if the self-duality of a given circuit is preserved during logic optimization and technology mapping. This chapter demonstrates that *Xor-majority graphs* (XMGs), as a graph representation during logic synthesis, can better abstract the Boolean property of self-duality. Using XMGs as the logic graph representation, the chapter proposes two strong Boolean optimizations of resubstitution and rewriting. The work demonstrates that using XMGs achieves better post technology-mapping area reduction as compared to the AIGs as demonstrated in Fig. 1.10. This is due

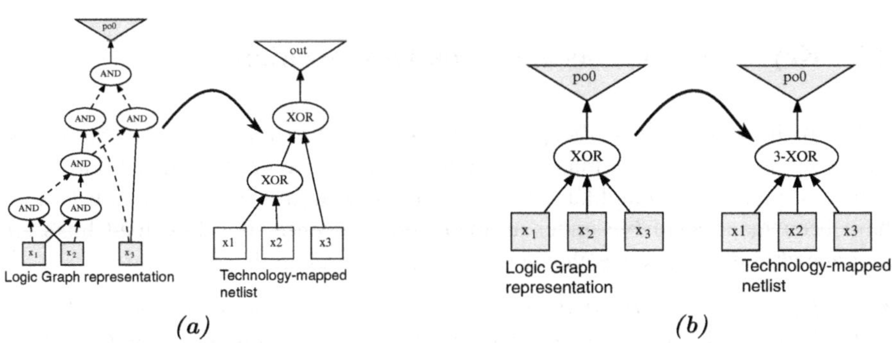

Fig. 1.10 Using different graph representations for mapping to different logic gates in case of RFETs. (**a**) AIG. (**b**) XMG

to the fact that since CMOS favors negative unate logic, AIG has been the natural choice of logic representation. Figure 1.10a shows how AIG favors two 2-input XORs as that is preferred in CMOS technology. In contrast, XMG favors mapping to a 3-input XOR which is preferred in RFET technology. This correlation with the logic synthesis has been shown in Fig. 1.9 via the green arrow that goes through the self-dual Boolean abstraction.

The next stage which is shown by the blue arrows is how the technology mapping connects with the physical synthesis flow. Physical synthesis uses standard cells based on RFETs to develop the final layout of the circuit based on RFETs. For this purpose, a technology model of SiNW RFET technology is considered and the overall flow of physical synthesis is presented in Chap. 6. The technology model is further used to develop `.lef` and `.lib` files. Both open-source and industrial tool flows are presented to carry out physical synthesis flow for RFET-based circuits. Further, two new approaches have been presented to get area improvements for a logic locking scenario depicting a secure circuit based on RFETs.

Once the EDA flow is defined, it can be used to develop secure circuits, which is the main application scenario for RFET-based circuits. Securing circuits based on RFETs through IP protection measures have been designed and analyzed in Chap. 7. While few of the prior works have focused on exploring security schemes using RFETs, a detailed evaluation with respect to the state-of-the-art security techniques has been missing. This chapter focuses on evaluating the security promises offered by RFET-based circuits. Not only security promises, but a disruptive security vulnerability has also been demonstrated, whose novelty and severity lie in the fact that they can be readily realized in an actual on-field RFET-based chip, either as an adversarial or a fail-safe measure. Both the security promises and vulnerabilities have been evaluated using a benchmark-level study to measure security on the basis of accepted metrics. Additionally, overheads, as compared to the CMOS technology, have been discussed and analyzed. Hardware security is synonymous to RFET-based circuits. That is why, in Fig. 1.9, hardware security is given in the background in red color as it encompasses the entire EDA flow and functionality-enhanced logic gates.

1.7 Key Contributions and Book Organization

In line with the research challenges and the specific questions presented earlier, the following are some of the major contributions that have been achieved during the course of this research and have been disseminated in this book. All contributions have been made available under an open-source license (the download links are mentioned in the respective chapters) to foster research in this direction.

- *Designs of combinational and sequential logic gates:* The book presents designs for both combinational and sequential logic gates. Both static and reconfigurable logic gates have been proposed. These designs are proposed in the work published in *IEEE Transactions on Very Large Scale Integration Systems*

(TVLSI)-2019 [Rai+19], *Proceedings of International Conference on Computer-Aided Design* (ICCAD)-2018 [Rai+18b], and *International Conference on Very Large Scale Integration* (VLSI-SOC)-2021 [Bha+21].
- *Identifying self-duality as the Boolean abstraction:* A Boolean abstraction is imperative to define specific EDA approaches and achieve improvements over the contemporary flows. The book identifies that the duality of functionality at the transistor level can be suitably abstracted using the Boolean property of self-duality. The methodology to distill standard cells using the property of self-duality is published in *Proceedings of Design Automation and Test in Europe conference* (DATE)-2020 [Rai+20a].
- *Technology mapping using the concept of HOF:* The book presents the concept of HOF to encapsulate reconfigurability at the logic level. This is essential to carry out an early-level evaluation of technology mapping using mutually exclusive multifunction logic gates. This is based on the work published in DATE-2018 [RRK18].
- *XMG-based logic synthesis*: Preserving self-duality at the circuit level is essential so that the maximum available self-dual portions of the logic network are mapped using self-dual logic gates. Hence, the book presents an XMG-based logic synthesis scheme that can preserve and utilize the existing self-duality to achieve better area reduction for RFET-based circuits as compared to the conventional logic synthesis flow. This is based on the work published in DATE-2021 [Rai+21c] and at *International Workshop on Logic and Synthesis* (IWLS) 2020 [Rai+20c]. An extension of the work [Rai+21c] is under revision at *IEEE Transactions on Computer-Aided Design of Integrated Circuits and Systems* (TCAD).
- *Physical synthesis flow:* The book presents an entire flow to carry out physical synthesis flow from a technology mapped netlist to the final .GDSII file. Library files such as the `.lef` and `.lib` files are developed to enable both an open-source and an industrial flow. The flow is based on the work published in DATE-2018 [Rai+18a] and ICCAD-2021 [Kri+21].
- *Security promises and vulnerabilities*: The book proposes both security promises and vulnerabilities for circuits based on RFETs. Security promises such as logic locking and split manufacturing have been proposed with RFET-based circuits. Evaluations demonstrate practical security with RFET-based circuits. Security vulnerabilities are demonstrated in sequential as well as combinational circuits. This work is published in *IEEE Transactions on Emerging Topics in Computing* (TETC)-2020 [Rai+20b] and DATE-2021 [Rai+21a].

A summary of the contributions in the entire book is presented in Table 1.1. It correlates the main title of the book and how it is broken down into topics that are covered in various chapters. It also correlates with the publications associated with individual chapters and also showcases which aforementioned challenges are tackled in which chapter.

The rest of the book is organized as follows: Chap. 2 presents and explains the relevant preliminaries for this book. It introduces different types of reconfigurable

Table 1.1 Correspondence between topics, chapters, publications, and challenges

	Topics	Chapters	Reference	Abstracting Ambipolarity	Enabling EDA	Hardware Security
Design Automation	Design Topologies for RFETs-based logic gates	Chapter 3	TCAD'19 [Rai+19], VLSI-SOC'21 [Bha+21] ICCAD'18 [Rai+18b]	●		
	Standard cells and enabling Technology Mapping	Chapter 4	DATE'18 [RRK18], DATE'20 [Rai+20a]	●	●	
	Logic Synthesis Flow	Chapter 5	DATE'21 [Rai+21c], IWLS'20 [Rai+20c]		●	●
	Physical Synthesis Flow	Chapter 6	DATE'18 [Rai+18a], ICCAD'21 [Kri+21]		●	●
Application	Hardware Security	Chapter 7	TETC'20 [Rai+20b], DSD'19 [RRK19], DATE'21 [Rai+21a]	●		●

technology such as the 1D and 2D devices. It further explains other terminologies related to design automation techniques. Then, Chap. 3 presents the handcrafted designs of both combinational and sequential logic gates. Evaluation in terms of area and delay has been shown in this chapter. It discusses how new design paradigms are relevant for RFET-based circuits as the conventional CMOS-based styles are suboptimal. The results of the Chap. 3 form the foundation for the next chapter as Chap. 4 investigates why a particular design style of logic gates can be efficiently built using RFETs. The chapter demonstrates how the Boolean property of self-duality can be efficiently abstracted using RFET-based logic gates. Chapter 4 also proposes an algorithm to distill logic functionalities that can be used in synthesis flow as standard cells by exploiting the property of self-duality found in the circuit descriptions. The chapter also introduces a new technology mapping that can utilize *mutually exclusive* logic functionalities during technology mapping. This is followed by Chap. 5 that proposes a logic synthesis flow that aims to preserve the existing self-duality in a given circuit. It is based on the premise of the previous chapter that self-dual logic gates can be efficiently implemented using RFETs. Within the context of logic synthesis, the chapter proposes various algorithms to further increase the existing self-duality to achieve further area reductions during technology mapping. Next, we have Chap. 6 that presents the library support for reconfigurable technologies to carry out the physical synthesis flow for RFET-based circuits. Improved layouts for static as well as reconfigurable logic gates are proposed with low pin-density to enable better placement and routing. Chapter 7 demonstrates how the transistor-level ambipolarity can be used to build efficient polymorphic logic gates based on RFETs. This enables security schemes such as logic locking and split manufacturing to be implemented with RFETs at lower overheads as compared to the contemporary CMOS technologies. Finally, concluding remarks and future work directions are given in Chap. 8.

References

[AGD14] Luca Amarú, Pierre-Emmanuel Gaillardon, and Giovanni De Micheli. "An Efficient Manipulation Package for Biconditional Binary Decision Diagrams". In: *Proceedings of the Conference on Design, Automation & Test in Europe*. DATE '14. Dresden, Germany: European Design and Automation Association, 2014, 296:1–296:6.

[AGM13] L. Amarù, P. E. Gaillardon, and G. De Micheli. "Efficient arithmetic logic gates using double-gate silicon nanowire FETs". In: *2013 IEEE 11th International New Circuits and Systems Conference (NEWCAS)*. June 2013, pp. 1–4. https://doi.org/10.1109/NEWCAS.2013.6573572.

[Ama+15a] L. Amarò, G. Hills, P. E. Gaillardon, S. Mitra, and G. De Micheli. "Multiple Independent Gate FETs: How many gates do we need?" In: *The 20th Asia and South Pacific Design Automation Conference*. IEEE, Jan. 2015, pp. 243–248. https://doi.org/10.1109/ASPDAC.2015.7059012.

[Ama+15b] L. Amarú, P. E. Gaillardon, S. Mitra, and G. De Micheli. "New Logic Synthesis as Nanotechnology Enabler". In: *Proceedings of the IEEE* 103.11 (Nov. 2015), pp. 2168–2195. https://doi.org/10.1109/JPROC.2015.2460377.

[Bas+21] Mohammad Khairul Bashar, Jaykumar Vaidya, R. S. Surya Kanthi, Chonghan Lee, Feng Shi, Vijaykrishnan Narayanan, and Nikhil Shukla. "Ferroelectric-Based Accelerators for Computationally Hard Problems". *In: Proceedings of the 2021 on Great Lakes Symposium on VLSI*. New York, NY, USA: Association for Computing Machinery, 2021, pp. 485–489.

[Bha+21] Abhiroop Bhattacharjee, Shubham Rai, Ansh Rupani, Michael Raitza, and Akash Kumar. "Metastability with Emerging Reconfigurable Transistors: Exploiting Ambipolarity for Throughput". In: *2021 IFIP/IEEE 29th International Conference on Very Large Scale Integration (VLSI-SoC)*. 2021, pp. 1–6. https://doi.org/10.1109/VLSI-SoC53125.2021.9607015.

[Bi+17] Yu Bi, Pierre-Emmanuel Gaillardon, X Sharon Hu, Michael Niemier, Jiann-Shiun Yuan, and Yier Jin. "Polarity-Controllable Silicon NanoWire FET-Based Security". In: *Security Opportunities in Nano Devices and Emerging Technologies*. CRC Press, 2017, pp. 165–178.

[BM10] Robert Brayton and Alan Mishchenko. "ABC: An academic industrialstrength verification tool". In: *Computer Aided Verification*. Springer. 2010.

[BM15] S. Bobba and G. De Micheli. "Layout Technique for Double-Gate Silicon Nanowire FETs With an Efficient Sea-of-Tiles Architecture". In: *IEEE Transactions on Very Large Scale Integration (VLSI) Systems* 23.10 (Oct. 2015), pp. 2103–2115. https://doi.org/10.1109/TVLSI.2014.2358884.

[Bor06] Shekhar Borkar. "Electronics beyond nano-scale CMOS". In: *Proceedings of the 43rd annual Design Automation Conference*. 2006, pp. 807–808.

[Bud+22] Ahmet F. Budak, Zixuan Jiang, Keren Zhu, Azalia Mirhoseini, Anna Goldie, and David Z. Pan. "Reinforcement Learning for Electronic Design Automation: Case Studies and Perspectives: (Invited Paper)". In: *2022 27th Asia and South Pacific Design Automation Conference (ASP-DAC)*. 2022, pp. 500–505. https://doi.org/10.1109/ASP-DAC52403.2022.9712578.

[BY17] Mark T Bohr and Ian A Young. "CMOS scaling trends and beyond". In: *IEEE Micro* 37.6 (2017), pp. 20–29.

[Cha+20] A. Chakraborty, N. G. Jayasankaran, Y. Liu, J. Rajendran, O. Sinanoglu, A. Srivastava, Y. Xie, M. Yasin, and M. Zuzak. "Keynote: A Disquisition on Logic Locking". In: *IEEE Transactions on Computer-Aided Design of Integrated Circuits and Systems* 39.10 (2020), pp. 1952–1972. https://doi.org/10.1109/TCAD.2019.2944586.

[CWD99] Jason Cong, Chang Wu, and Yuzheng Ding. "Cut ranking and pruning: Enabling a general and efficient FPGA mapping solution". In: *Proceedings of the 1999 ACM/SIGDA seventh international symposium on Field programmable gate arrays*. 1999, pp. 29–35.

[Den+74] R.H. Dennard, F.H. Gaensslen, Hwa-Nien Yu, V.L. Rideout, E. Bassous, and A.R. LeBlanc. "Design of ion-implanted MOSFET's with very small physical dimensions". In: IEEE Journal of Solid-State Circuits 9.5 (1974), pp. 256–268. https://doi.org/10.1109/JSSC.1974.1050511.

[DW04] Andre DeHon and Michael J. Wilson. "Nanowire-based Sublithographic Programmable Logic Arrays". In: *Proceedings of the 2004 ACM/SIGDA 12th International Symposium on Field Programmable Gate Arrays*. FPGA '04. Monterey, California, USA: ACM, 2004, pp. 123–132. https://doi.org/10.1145/968280.968299.

[Gai+13] Pierre-Emmanuel Gaillardon, Michele De Marchi, Luca Amarù, Shashikanth Bobba, Davide Sacchetto, Yusuf Leblebici, and Giovanni De Micheli. "Towards structured ASICs using polarity-tunable Si nanowire transistors". In: *Proceedings of the 50th Annual Design Automation Conference on - DAC '13* (2013), p. 1. https://doi.org/10.1145/2463209.2488886.

[Gai+15] Pierre Emmanuel Gaillardon, Xifan Tang, Gain Kim, and Giovanni De Micheli. "A Novel FPGA Architecture Based on Ultrafine Grain Reconfigurable Logic Cells". In: *IEEE Transactions on Very Large Scale Integration (VLSI) Systems* 23.10 (Oct. 2015), pp. 2187–2197. https://doi.org/10.1109/TVLSI.2014.2359385.

[GAM14] P. E. Gaillardon, L. G. Amarù, and G. D. Micheli. "Unlocking Controllable-Polarity Transistors Opportunities by Exclusive-OR and Majority Logic Synthesis". In: *2014 IEEE Computer Society Annual Symposium on* VLSI. July 2014, pp. 403–405. https://doi.org/10.1109/ISVLSI.2014.107.

[GMT22] Giulio Galderisi, Thomas Mikolajick, and Jens Trommer. "Reconfigurable Field Effect Transistors Design Solutions for Delay-Invariant Logic Gates". In: *IEEE Embedded Systems Letters* (2022), pp. 1–1. https://doi.org/10.1109/LES.2022.3144010.

[Hei+13] André Heinzig, Thomas Mikolajick, Jens Trommer, Daniel Grimm, and Walter M. Weber. "Dually Active Silicon Nanowire Transistors and Circuits with Equal Electron and Hole Transport". In: *Nano Letters* 13.9 (2013). PMID: 23919720, pp. 4176–4181. https://doi.org/10.1021/nl401826u. eprint: https://dx.doi.org/10.1021/nl401826u.

[Hil+19] G. Hills, C. Lau, A. Wright, S. Fuller, M. D Bishop, T. Srimani, P. Kanhaiya, R. Ho, A Amer, Y Stein, et al. "Modern microprocessor built from complementary carbon nanotube transistors". In: *Nature* (2019).

[KJJ99] Paul Kocher, Joshua Jaffe, and Benjamin Jun. "Differential Power Analysis". In: *Advances in Cryptology*. Springer. 1999, pp. 388–397.

[Kne20] Johann Knechtel. "Hardware Security For and Beyond CMOS Technology: An Overview on Fundamentals, Applications, and Challenges". In: *Proceedings of the 2020 International Symposium on Physical Design*. New York, NY, USA: Association for Computing Machinery, 2020, pp. 75–86.

[Kri+21] Andreas Krinke, Shubham Rai, Akash Kumar, and Jens Lienig. "Exploring Physical Synthesis for Circuits based on Emerging Reconfigurable Nanotechnologies". In: *2021 IEEE/ACM International Conference On Computer Aided Design (ICCAD)*. IEEE. 2021, pp. 1–9.

[Mas+17] Mohamed El Massad, Jun Zhang, Siddharth Garg, and Mahesh V. Tripunitara. "Logic Locking for Secure Outsourced Chip Fabrication: A New Attack and Provably Secure Defense Mechanism". In: *CoRR* abs/1703.10187 (2017).

[MBV06] Valavan Manohararajah, Stephen Dean Brown, and Zvonko G Vranesic. "Heuristics for area minimization in LUT-based FPGA technology mapping". In: *IEEE Transactions on Computer-Aided Design of Integrated Circuits and Systems* 25.11 (2006), pp. 2331–2340.

[Mic94] Giovanni De Micheli. *Synthesis and optimization of digital circuits*. 1st. McGraw-Hill, 1994, p. 579.

[Mik+17] T Mikolajick, A Heinzig, J Trommer, T Baldauf, and W M Weber. "The RFET—a reconfigurable nanowire transistor and its application to novel electronic circuits and systems". In: *Semiconductor Science and Technology* 32.4 (2017), p. 043001.

[Mik+21] T Mikolajick, G Galderisi, M Simon, S Rai, A Kumar, A Heinzig, WM Weber, and J Trommer. "20 Years of reconfigurable field-effect transistors: From concepts to future applications". In: *Solid-State Electronics* 186 (2021), p. 108036.

[Moo+65] Gordon E Moore et al. *Cramming more components onto integrated circuits*. 1965.

[Pan18] Manish Pandey. "Machine learning and systems for building the next generation of EDA tools". In: *2018 23rd Asia and South Pacific Design Automation Conference* (ASP-DAC). 2018, pp. 411–415. https://doi.org/10.1109/ASPDAC.2018.8297358.

[Per+21] Yasasvi V. Peruvemba, Shubham Rai, Kapil Ahuja, and Akash Kumar. "RL-Guided Runtime-Constrained Heuristic Exploration for Logic Synthesis". In: *2021 IEEE/ACM International Conference On Computer Aided Design (ICCAD)*. 2021, pp. 1–9. https://doi.org/10.1109/ICCAD51958.2021.9643530.

[Rad+] I P Radu, O Zografos, A Vaysset, J Yan, J Swerts, D Radisic, B Soree, M Manfrini, M Ercken, P Raghavan, S Sayan, C Adelmann, and A Thean. "Spintronic majority gates". In: *()*, pp. 3–6.

[Rai+17] M. Raitza, A. Kumar, M. Völp, D.Walter, J. Trommer, T. Mikolajick, and W. M. Weber. "Exploiting transistor-level reconfiguration to optimize combinational circuits". In: *Design, Automation Test in Europe Conference Exhibition (DATE), 2017. Mar. 2017*, pp. 338–343. https://doi.org/10.23919/DATE.2017.7927013.

[Rai+18a] S. Rai, A. Rupani, D. Walter, M. Raitza, A. Heinzig, T. Baldauf, J. Trommer, C. Mayr, W. M. Weber, and A. Kumar. "A physical synthesis flow for early technology evaluation of silicon nanowire based reconfigurable FETs". In: *2018 Design, Automation Test in Europe Conference Exhibition (DATE)*. Mar. 2018, pp. 605–608. https://doi.org/10.23919/DATE.2018.8342080.

[Rai+18b] Shubham Rai, Srivatsa Srinivasa, Patsy Cadareanu, Xunzhao Yin, Xiaobo Sharon Hu, Pierre-Emmanuel Gaillardon, Vijaykrishnan Narayanan, and Akash Kumar. "Emerging Reconfigurable Nanotechnologies: Can They Support Future Electronics?" In: *Proceedings of the International Conference on Computer-Aided Design*. ICCAD '18. San Diego, California: ACM, 2018, 13:1–13:8. https://doi.org/10.1145/3240765.3243472.

[Rai+19] S. Rai, J. Trommer, M. Raitza, T. Mikolajick, W. M. Weber, and A. Kumar. "Designing Efficient Circuits Based on Runtime-Reconfigurable Field-Effect Transistors". In: *IEEE Transactions on Very Large Scale Integration (VLSI) Systems* 27.3 (Mar. 2019), pp. 560–572. https://doi.org/10.1109/TVLSI.2018.2884646.

[Rai+20a] S. Rai, M. Raitza, S.S. Sahoo, and A. Kumar. "DiSCERN: Distilling Standard-Cells for Emerging Reconfigurable Nanotechnologies". In: *DATE*. 2020.

[Rai+20b] Shubham Rai, Satwik Patnaik, Ansh Rupani, Johann Knechtel, Ozgur Sinanoglu, and Akash Kumar. "Security Promises and Vulnerabilities in Emerging Reconfigurable Nanotechnology-Based Circuits". In: *IEEE Transactions on Emerging Topics in Computing* (2020), pp. 1–1. https://doi.org/10.1109/TETC.2020.3039375.

[Rai+20c] Shubham Rai, Heinz Riener, Giovanni De Micheli, and Akash Kumar. "XMG-based Logic Synthesis for Emerging Reconfigurable Nanotechnologies". In: *[Proceedings of the 29th International Workshop on Logic & Synthesis (IWLS 2020)]*. CONF. 2020.

[Rai+21a] Shubham Rai, Mengyun Liu, Anteneh Gebregiorgis, Debjyoti Bhattacharjee, Krishnendu Chakrabarty, Said Hamdioui, Anupam Chattopadhyay, Jens Trommer, and Akash Kumar. "Perspectives on Emerging Computation-in-Memory Paradigms". In: *2021 Design, Automation Test in Europe Conference Exhibition (DATE)*. 2021, pp. 1925–1934. https://doi.org/10.23919/DATE51398.2021.9473976.

[Rai+21b] Shubham Rai, Walter Lau Neto, Yukio Miyasaka, Xinpei Zhang, Mingfei Yu, Qingyang Yi, Masahiro Fujita, Guilherme B. Manske, Matheus F. Pontes, Leomar S. da Rosa, Marilton S. de Aguiar, Paulo F. Butzen, Po-Chun Chien, Yu-Shan Huang, Hoa-Ren Wang, Jie-Hong R. Jiang, Jiaqi Gu, Zheng Zhao, Zixuan Jiang, David Z. Pan, Brunno A. de Abreu, Isac de Souza Campos, Augusto Berndt, Cristina Meinhardt, Jonata T. Carvalho, Mateus Grellert, Sergio Bampi, Aditya Lohana, Akash Kumar, Wei Zeng, Azadeh Davoodi, Rasit O. Topaloglu, Yuan Zhou, Jordan Dotzel,

Yichi Zhang, Hanyu Wang, Zhiru Zhang, Valerio Tenace, Pierre-Emmanuel Gaillardon, Alan Mishchenko, and Satrajit Chatterjee. "Logic Synthesis Meets Machine Learning: Trading Exactness for Generalization". In: *2021 Design, Automation Test in Europe Conference Exhibition (DATE)*. 2021, pp. 1026–1031. https://doi.org/10.23919/DATE51398.2021.9473972.

[Rai+21c] Shubham Rai, Heinz Riener, Giovanni De Micheli, and Akash Kumar. "Preserving Self-Duality During Logic Synthesis for Emerging Reconfigurable Nanotechnologies". In: *2021 Design, Automation Test in Europe Conference Exhibition (DATE)*. 2021, pp. 354–359. https://doi.org/10.23919/DATE51398.2021.9474112.

[Raj+15] J. Rajendran, R. Karri, J. B. Wendt, M. Potkonjak, N. McDonald, G. S. Rose, and B. Wysocki. "Nano Meets Security: Exploring Nanoelectronic Devices for Security Applications". In: *Proceedings of the IEEE* 103.5 (May 2015), pp. 829–849. https://doi.org/10.1109/JPROC.2014.2387353.

[Reu+21] Maximilian Reuter, Johannes Pfau, Tillmann A. Krauss, Jülrgen Becker, and Klaus Hofmann. "From MOSFETs to Ambipolar Transistors: Standard Cell Synthesis for the Planar RFET Technology". In: *IEEE Transactions on Circuits and Systems I: Regular Papers* 68.1 (2021), pp. 114–125. https://doi.org/10.1109/TCSI.2020.3035889.

[RRK18] S. Rai, M. Raitza, and A. Kumar. "Technology mapping flow for emerging reconfigurable silicon nanowire transistors". In: *2018 Design, Automation Test in Europe Conference Exhibition (DATE)*. Mar. 2018, pp. 767–772. https://doi.org/10.23919/DATE.2018.8342110.

[RRK19] A. Rupani, S. Rai, and A. Kumar. "Exploiting Emerging Reconfigurable Technologies for Secure Devices". In: *Euromicro* DSD. 2019.

[Sha20] John Shalf. "The future of computing beyond Moore's law". In: *Philosophical Transactions of the Royal Society A* 378.2166 (2020), p. 20190061.

[Sim+16] M. Simon, A. Heinzig, J. Trommer, T. Baldauf, T. Mikolajick, and W. M. Weber. "Bringing reconfigurable nanowire FETs to a logic circuits compatible process platform". In: *2016 IEEE Nanotechnology Materials and Devices Conference (NMDC)*. Oct. 2016, pp. 1–3. https://doi.org/10.1109/NMDC.2016.7777085.

[Sim+18] M. Simon, J. Trommer, B. Liang, D. Fischer, T. Baldauf, M. B. Khan, A. Heinzig, M. Knaut, Y. M. Georgiev, A. Erbe, J. W. Bartha, T. Mikolaiick, and W. M. Weber. "A wired-AND transistor: Polarity controllable FET with multiple inputs". In: *2018 76th Device Research Conference (DRC)*. June 2018, pp. 1–2. https://doi.org/10.1109/DRC.2018.8442159.

[Tro+15] Jens Trommer, André Heinzig, Tim Baldauf, Stefan Slesazeck, Thomas Mikolajick, and Walter M. Weber. "Functionality-Enhanced Logic Gate Design Enabled by Symmetrical Reconfigurable Silicon Nanowire Transistors". In: *IEEE Transactions on Nanotechnology* 14.4 (July 2015), pp. 689–698. https://doi.org/10.1109/TNANO.2015.2429893.

[Tro+16] J. Trommer, A. Heinzig, T. Baldauf, T. Mikolajick, W. M. Weber, M. Raitza, and M. Völp. "Reconfigurable nanowire transistors with multiple independent gates for efficient and programmable combinational circuits". In: *2016 Design, Automation Test in Europe Conference Exhibition (DATE)*. Mar. 2016, pp. 169–174.

[YC16] Shimeng Yu and Pai-Yu Chen. "Emerging Memory Technologies: Recent Trends and Prospects". In: *IEEE Solid-State Circuits Magazine* 8.2 (2016), pp. 43–56. https://doi.org/10.1109/MSSC.2016.2546199.

[Zha+14] Jian Zhang, Xifan Tang, Pierre Emmanuel Gaillardon, and Giovanni De Micheli. "Configurable Circuits Featuring Dual-Threshold-Voltage Design With Three-Independent-Gate Silicon Nanowire FETs". In: *IEEE Transactions on Circuits and Systems I: Regular Papers* 61.10 (Oct. 2014), pp. 2851–2861. https://doi.org/10.1109/TCSI.2014.2333675.

Chapter 2
Preliminaries

The previous chapter presented an introduction to the emerging reconfigurable nanotechnologies and the related challenges associated with enabling circuits based on such technology. This chapter is dedicated to explaining the background and preliminaries needed throughout the whole book. The chapter can be broadly classified into two main parts. The first part covers different types of reconfigurable nanotechnology such as 1D and 2D devices. Then, the device physics of nanowire-based RFETs is described to explain transistor-level dynamic reconfiguration in such technology. Feasibility aspects in terms of commercial adoption for reconfigurable nanotechnology are also presented.

The second part covers the background of design automation techniques. Particularly, concepts such as monotone functions are explained. Important terminologies such as graph representation to carry out logic synthesis and technology mapping are explained.

2.1 Reconfigurable Nanotechnology

Reconfigurable technology refers to an emerging class of nanotechnologies in which the devices (or transistors) exhibit electrical conduction for both types of charge carriers—electrons or holes, on the application of an external bias potential [Mar+12, Hei+12]. These devices are referred as *reconfigurable field-effect transistors* (RFETs) or *polarity-control* devices. This duality in electrical conduction is independently (re)programmable, and hence, the devices can exhibit either p- or n-type functionality [Mik+17]. This reconfigurability is not enabled via chemical doping due to material engineering or impurities but through electrostatic doping, i.e., the generation of charge carriers through the channel via an external

S. Rai, A. Kumar, *Design Automation and Applications for Emerging Reconfigurable Nanotechnologies*, https://doi.org/10.1007/978-3-031-37924-6_2

potential. This device-level[1] reconfiguration is due to the phenomenon called *ambipolarity* or ambipolar conduction.

Ambipolarity is a phenomenon observed at 22-nm or lower technology nodes that allows conduction of both charge carriers through the transistor channel. During device engineering for conventional transistors below 22 nm, ambipolarity is a potential limitation in circuit design and is often suppressed [Mar+14]. However, ambipolar conduction is enhanced using process techniques for several nanoscale FET devices based on various low gap materials such as silicon or germanium [Mar+12]. The enhancement of ambipolarity enables symmetric electric conduction in both p- and n-type configurations. This reconfiguration of charge carriers through the channel is also termed as *polarity control* [Mar+12].

In terms of the structure of the FET and the physical properties, reconfiguration in an RFET can be broadly classified into two main types—Schottky-barrier Field-Effect Transistor (SBFET) and band to band Tunneling Field-Effect Transistors (TFETs). In case of SBFETs, the flow of charge carriers is controlled through a Schottky-Barrier (SB) at the metal-semiconductor junction. The type and injection of the charge carriers are enabled by independent gating of each of the two junctions. This can be done in two ways—direct control of the carrier injection using a potential bias at the Schottky junctions [Hei+12] or through control of the transport of previously injected charge carriers using a charge barrier [Mar+12]. In the former case, the control is done through the potential over the Schottky barrier itself, while in the latter case, the injection of the charge carrier is allowed through two additional contacts above (or below) the Schottky junctions [Mik+21]. By applying the potential at the two polarity gate contacts, the type of charge carriers injected can be controlled, while the conduction is modulated through a potential applied at the middle of the channel. Both these scenarios lead to bending of the energy bands in the semiconductor and the concept can be referred to as *electrostatic doping*. However, as compared to the former case, the electrostatic doping developed in the second case offers a higher electrical symmetry because the two types of gate terminals do not compete with each other regarding the control of a certain channel region. Within RFETs, the direct control of the charge injection is demonstrated in Dual-independent gate Field-Effect Transistor (DIGFET) as proposed in [Hei+12]. RFETs with two additional polarity control is demonstrated in devices with *gate-all-around* structures as shown in TIGFETs [Mar+12, Gor+19, Tro+17b].

Similarly, reconfigurable properties in case of TFET [KM14, MR16] is demonstrated using Band-to-Band Tunneling (BTBT) as the conduction mechanism. Here, chemical doping is used to create two highly doped silicon regions separated by the nearly intrinsic channel region. By applying potential at the two contact regions, it is possible to induce enough band bending to allow charge carriers to flow. However, due to a high doping intensity, TFET devices are generally slow and suffer through low on-currents.

[1] The term device-level and transistor-level are used interchangeably throughout this book.

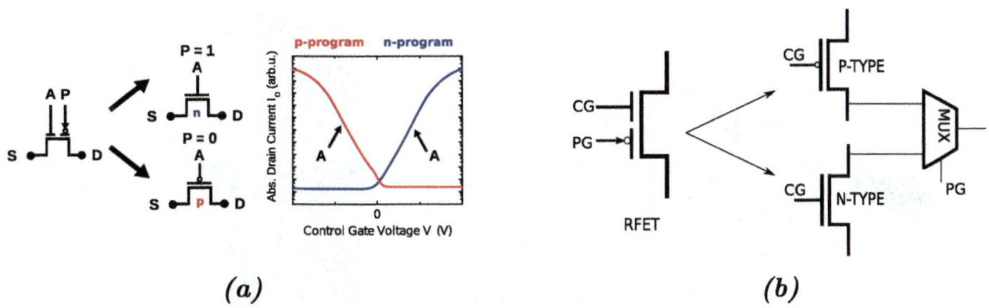

Fig. 2.1 (**a**) A schematic representation of an RFET showing two gate terminals: the program (signal P) and the CG (signal A) [Rai+19]. The PG controls the type of charge carriers, whereas the CG controls the flow of the charge carriers. The adjacent curve shows the V-shaped curve representing electrical symmetry for n- and p-type functionality. (**b**) From a logical abstraction point of view, RFET comparison with CMOS

The devices belonging to reconfigurable nanotechnologies are characterized by two types of gate terminals—the *program gate* (PG) and the *control gate* (CG). A representative RFET with all types of terminals is shown in Fig. 2.1a. These two gate terminals can be independently controlled and can alter the conduction through the channel. Potential bias at the PG terminal can electrically block a particular type of charge carrier flowing through the channel. Hence, the PG controls the electrostatic behavior of the device to function either as a p- or n-type device. The CG controls the flow of the charge carriers by allowing charge carriers to accumulate within the channel. The CG is similar to the gate terminal in typical CMOS technology.

The unique characteristic of reconfigurable nanotechnology is the ability to demonstrate electrical symmetry in both p- and n-type configurations. The current-voltage curve of these transistors is shown in Fig. 2.1a. In terms of logical abstraction with reference to the contemporary CMOS, an RFET can be represented as shown in Fig. 2.1b. It can be seen that the select line of the multiplexer is connected to the PG terminal and that controls the functioning of the device as a p- or an n-type device.

In terms of the available geometry, RFETs can be broadly classified into two main types—1D devices and 2D devices. Next, these concepts are explained in detail.

2.1.1 1D Devices

One-dimensional devices refer to those RFETs that come in one-dimensional geometry such as Silicon Nanowire (SiNW) [Hei+12, Mar+12] or Germanium Nanowire (GeNW) [Tro+17a], CNTFETs [Lin+05], or graphene nanoribbons [Har+10]. The main focus of this book is the nanowire-based RFETs.

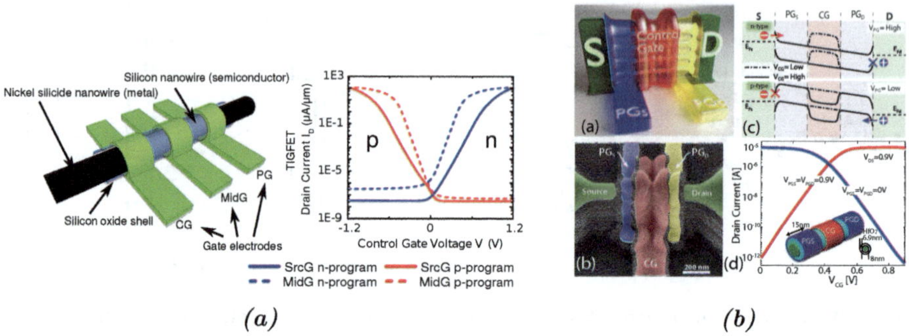

Fig. 2.2 (**a**) A representation of a nanowire-based RFETs with three all-around gate structure [Rai+17]. The right figure shows the I–V characteristics. (**b**) It shows an all-around RFET, called *three-independent-gate FETs* (TIGFETs). The band diagrams are shown from [Rai+18b]

Figure 2.2a represents a 1D dopant-free monocrystalline nanowire structure made of silicon as shown in [Rai+17]. The nickel silicide metal contacts form Schottky junctions at the source and drain contacts. The gate overlapping with the source is the Control Gate (CG), while the gate overlapping with the drain is the Program Gate (PG) as shown in Fig. 2.2b [Gai+13b]. Steering these two gate terminals controls the polarity of the device. Transfer characteristics represent fully symmetrical p-type and n-type functionality [Hei+12]. The device can be steered either with three or two gates. In the three-gate configuration with the middle-gate MidG as the CG, the dashed transfer curves are obtained. The signal at the outer gates, DrnG and SrcG, control the p-type or n-type nature of the transistor. In the dual-gate configuration, the middle gate is left out in the design and the solid transfer curves are obtained. In the latter, DrnG acts as the PG selecting p-type or n-type configuration and SrcG acts as the CG.

Just like silicon, germanium nanowire-based RFETs have also been demonstrated. Germanium being a low bandgap material is a promising channel material due to its integrability with the CMOS manufacturing process [Tro+14]. Being a low-band channel material, GeNW offers higher static currents as compared to SiNW-based RFETs. Hence, as compared to silicon, GeNW-based RFETs offers better current drives. Conversely, they also suffer due to high leakage dissipation.

Working Principle of a Nanowire-Based RFET

An RFET is a multigate structure containing two types of gate terminals. Here, for the sake of understanding, a DIGFET is considered. The junction contacts at the source and drain are Schottky contacts [Tro+15]. Figure 2.3 shows the polarity control due to the bending of the energy band in RFETs. In the off state, the current is shut off, due to the barrier induced by the opposing potential at the PG and the CG. In the on state of the n-type (or the p-type), the CG enables electron

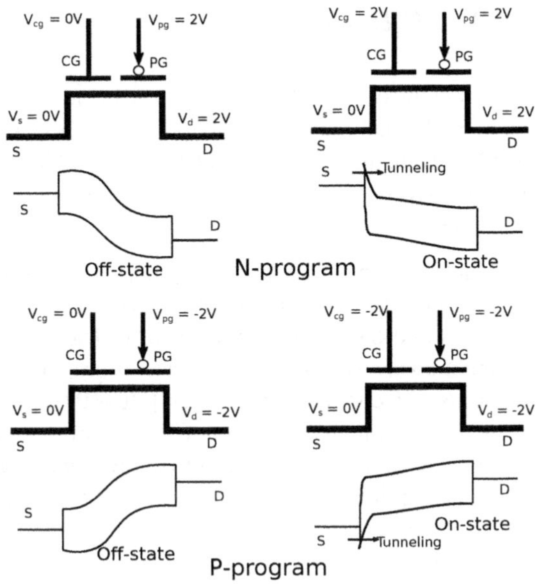

Fig. 2.3 Conceptual representation of a working principle for a nanowire-based RFET. One can notice how the bands move on application of potential at PG and CG. The on state is $|V_{CG}| = 2$ V and $|V_{CG}| = 0$ V for n- and p-type operation, respectively [Tro+15]

Table 2.1 Programming RFETs [Tro+15]

Functionality	Potential at PG V_{pg}	Potential at source V_S	Potential at drain V_D
n-FET	DD	Low	High
p-FET	GND	High	Low

tunneling (hole tunneling) through the Schottky junctions by bending down the silicon bands as shown in Fig. 2.3. For Gate-All Around (GAA)-based RFETs as proposed in [Mar+12, Zha+14b], the potential at the CG and the all-around PGs bend the silicon band in a similar way, which allows tunneling currents based on majority carriers through the Schottky junctions. However, since they have two PG on either side of the control gate, they demonstrate a wider range of operating stages. By changing the potential on either side of CG, the range of operating stages can alter the performance of RFET. Further details regarding the physics of such reconfigurable devices can be found in [Mik+17, Zha+14b].

At the logical abstraction level, an RFET is a programmable device that can be tuned to specific electrical behavior depending upon the PG, source, and drain potential. This is shown in Table 2.1. With the default configuration shown in Table 2.1, the device is *ON* in n-FET (p-FET) configuration when CG is at VDD (GND) and vice versa. This runtime reconfigurability, in turn, leads to functional flexibility at the logic gate level, where a single logic gate exhibits more than one functionality [Rai+19].

2.1.2 2D Devices

Transistor-level reconfigurability has been demonstrated in many planar devices made of channel materials, such as silicon [Reu+21], graphene [Tan+10] [Mir+13], or other Transition Metal Dicalchogenide (TMD) materials like $MoTe_2$ [Nak+15] or WSe_2 [Res+16]. The following are the major 2D devices that have been under active research:

1. Graphene p-n junction: A graphene p-n junction as shown in Fig. 2.4a consists of a graphene sheet with two metal-to-graphene contacts; A and Z are the inputs and the outputs, respectively. The two sheets are separated using a thick oxide layer. Ambipolarity in case of graphene p-n junctions (Fig. 2.4a) is due to the use of coplanar split gates [Hua+07], which are similar to different types of gate terminals (PG and CG) as present in SiNW RFETs. Their extreme thinness offers superior electrostatic control and they are conductive for low-power applications. The Verilog-A model of graphene p-n junction has also been proposed in [Mir+13]. Reconfigurable properties for graphene p-n junctions can be seen in Fig. 2.4b.
2. TMD devices: Tungsten diselenide (WSe_2) is one of the stable 2D TMD that can demonstrate high performance. Exemplary electrical symmetry can be seen

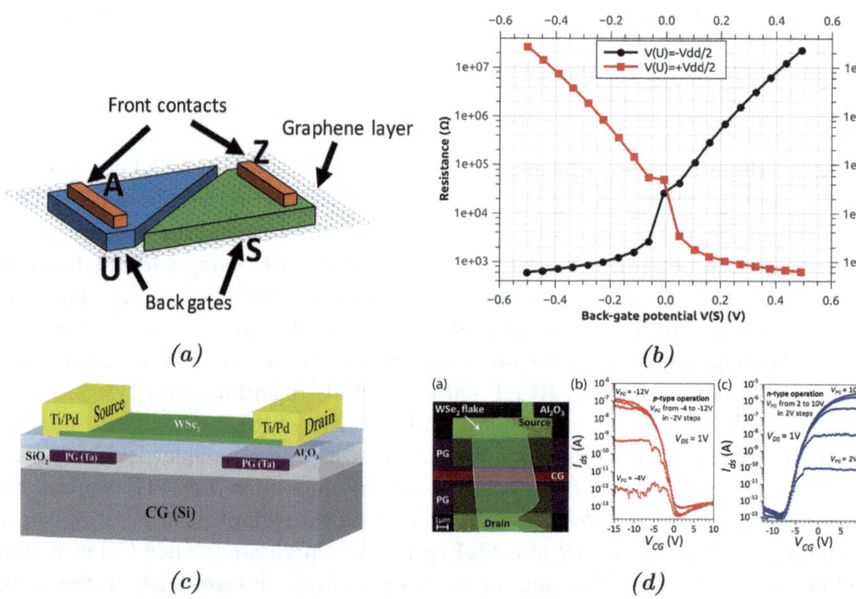

Fig. 2.4 (**a**) A graphene p-n junction is shown where the back gates (S and U) work as a control knob to control the ambipolarity. (**b**) Resistance variation vs back gate potential [Ten+18]. (**c**) Cross section of the WSe_2 device [Res+16]. (**d**) Demonstration of WSe_2 TIGFET along with the transfer characteristics [Res+16]

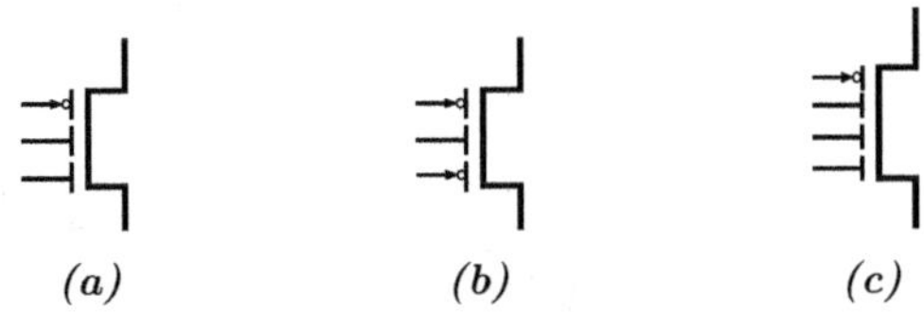

Fig. 2.5 RFETs with multiple independent gate terminals. (**a**) TIGFET. (**b**) GAA TIGFET. (**c**) MIGFET

for both p- and n-type functionality. Experimental evaluations demonstrate that WSe_2 can be a promising candidate to develop planar TIGFETs that can realize high performance circuits. The fabricated device is shown in Fig. 2.4c. Another RFET material that has been demonstrated is the molybdenum ditelluride ($MoTe_2$).

2.1.3 Factors Favoring Circuit Flexibility

Apart from transistor-level reconfigurability, RFETs offer exceptional properties that allow higher flexibility for circuit design as compared to the conventional CMOS technology.

Multiple Independent Gate Terminals

Since the conduction in nanowire channels is due to tunneling of charge carriers, it has been shown that within the channel, the on-current is only dependent upon the injection at the Schottky barriers. Consequently, this allows the RFET to operate with an ungated area in the middle of the channel. This has been demonstrated through measurements [Hei+12] and simulations [Tro+15]. This particular feature has even been both theoretically proven and experimentally shown by Trommer et al. [Tro+16] and Zhang et al. [Zha+14a], respectively. This provides an opportunity to add multiple independent gate terminals on the channel. Further, as the on resistance of the device is dominated by the resistance of the source-sided barrier, there is nearly no degradation in the current through the device by increasing the channel length, as long as good electrostatic control over the channel potential is maintained. With multiple, independent gate transistors, there is just a slight increase in the overall channel resistance of the multigated device, owing to the Schottky barrier in the on state of an RFET device [Tro+16]. This can be seen in Fig. 2.6 where the three inputs operate at a virtually lower resistance than the internal resistance of the device.

The various schematic of different types of RFETs—TIGFETs, GAAs TIGFETs, and Multi-independent gate Field-Effect Transistors (MIGFETs)—are shown in Fig. 2.5a,b, and c, respectively. Throughout the manuscript, unless otherwise specified, an RFET with two gate terminals is used as a general representation for all types of RFETs.

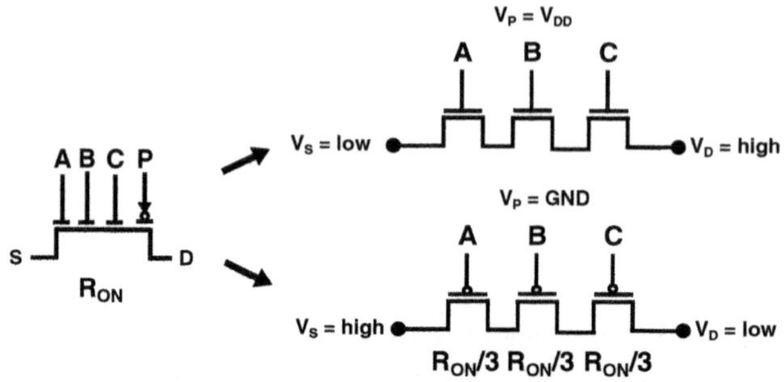

Fig. 2.6 Equivalent circuit of a multigate RFET with four gates. One single transistor resembles three single-gated transistors in series. All three transistors are programmed to p- or n-type functionality depending on the applied voltage scheme. In addition, they virtually operate with an effective resistance of only $\frac{1}{3}$ of the internal resistance of the device [Rai+19]

The presence of multi-independent gate terminals on a single channel of a nanowire-based RFET eliminates the need for multi-level stacked transistors as is the case in the CMOS paradigm, thereby doing away with the individual load offered by the capacitance (e.g., gate-to-source capacitance) from individual transistors cascaded in series [Zha+14b]. Such functionality has been exploited in works like [Sim+18] to design a wired-AND transistor containing multiple independent logical inputs requiring a single supply voltage. These multi-independent gate terminals extend the device functionality because a single transistor can incorporate many inputs thereby giving rise to higher functionality with less number of transistors. These transistors facilitate the designing of logic gates with more than three inputs as shown in the next chapter (Fig. 3.1). These logic gate designs show better performance in terms of normalized delay as compared to CMOS designs.

Modifiable Performance

In a typical CMOS technology, the threshold voltage (V_t) defines the performance of the device. Devices with low V_t are referred to as high-performance devices, while devices with high V_t are referred to as low-power devices. The general convention is that at a particular technology node, both devices are available, and depending upon the requirements of the circuit design, a cocktail of different threshold voltage devices (*multi-*V_t) are used. In CMOS, various techniques such as altering the dopant concentration [Mat+09] and body biasing [Tsc+02] can be utilized to alter the threshold potential.

In case of GAA TIGFETs, the two program gate terminals can be utilized to alter the performance of a single device. This has been first demonstrated in [Zha+14a] and is shown in Fig. 2.7. If both the PG are biased to a single potential which is also equal to the bias applied to the CG, then the device operates in Low V_t (Threshold Voltage) (LVT) mode (shown in Fig. 2.7a and b). The device operates either in the

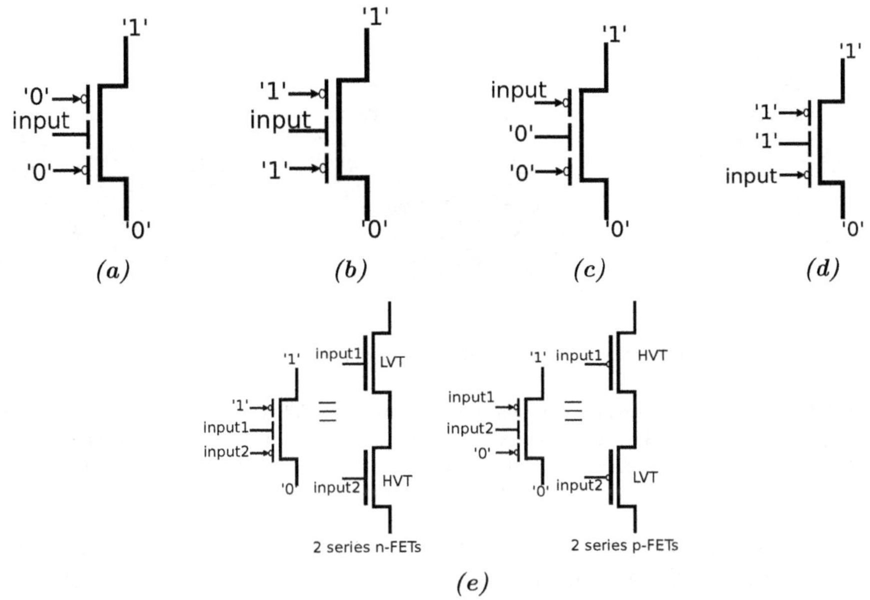

Fig. 2.7 Various operating modes in the case of Gate-All Around (GAA) TIGFET. (**a**) LVT PFET. (**b**) LVT NFET. (**c**) HVT PFET. (**d**) HVT NFET. (**e**) Series operation in case of GAA RFET

on state or in the off state. Conversely, if one PG and CG are biased with the same potential, leaving the other PG to be biased with the input, the device operates in High V_t (Threshold Voltage) (HVT) mode (shown in Fig. 2.7c and d). In this case, the device operates either in the on state or the low-leakage off state [Zha+14b].

By doing the connections as shown in Fig. 2.7e, GAA TIGFET operate similar to a DIGFET. This is possible due to the combination of LVT and HVT operating states. This is similar to the cascading of multiple transistors, albeit with different threshold voltages realizing a multi-V_t design.

2.2 Feasibility Aspects of RFET Technology

While there are several apprehensions about the feasibility aspects of emerging nanotechnologies, there are several reasons which motivate us to look at security features for circuits based on reconfigurable nanotechnology.

- Works like [Mik+17, Sim+17] have shown that reconfigurable nanotechnologies follow the similar fabrication and manufacturing process as CMOS. Even the proposed 2D devices such as graphene p-n junctions can be fabricated with minimal patterning and hence are compatible with CMOS fabrication schemes.

(a) Planar FET geometry *(b) FinFET geometry* *(c) Nanosheet geometry*

Fig. 2.8 Evolution of geometry for field-effect transistors [YEK19]. (**a**) Planar FET designs were popular till 2011. (**b**) FinFET designs were proposed to further reduce the channel width. This covers the channel region from three sides. (**c**) Stacked nanosheet designs where the gate completely surrounds the channel region to give better control than FinFET

- In terms of geometries of these newer nanotechnologies, Si and Ge nanowire geometries are the natural successor for FinFET-based transistors, which is the main fraction of modern CMOS technologies [Cou16]. Works from both the industry and academia show that a FinFET can be replaced efficiently by stacked nanowires [Chu+18, Lou+17]. This can be seen from Fig. 2.8. It can be seen from the figure that prior to 2011, planar geometry was favored which was replaced by FinFETs for even larger-scale integration. However, to further push for the device performance, 1D (as nanowire geometry) and 2D (as sheet or planar geometry) are promising options.
- As we have seen for 1D and 2D devices, there is no dearth of reconfigurable nanotechnologies as they are realized with various materials like carbon [Lin+05], graphene [Har+10, Tan+10], germanium [Tro17], and WSE_2 [Res+16]. This is a clear indication that reconfigurability at the technology level has many contenders.
- Various works like [Gai+13a, Rai+18a, Rai+17] have demonstrated advanced circuit-level designs and corresponding EDA flows using RFETs with detailed evaluation. Further, RFETs allow multiple options for functional flexibility–either in terms of the number of gate terminals on a single channel [Zha+14b] or configuring the performance of a device [Zha+14a]. As we can see using HVT and LVT operations, both high performance, as well as low power devices, can be manufactured. Technologies such as WSe_2 have been demonstrated to be a suitable candidate for high-performance applications [Yu+15].
- Polymorphism at the transistor level is an important criterion for hardware security as it can inherently support both camouflaging and locking [Rai+20]. Security applications using such reconfigurable nanotechnologies have already been demonstrated in works like [Raj+15, Bi+16, Che+16, RRK19] by utilizing their unique I–V characteristic which is generally not possible with conventional MOSFETs.

- Several reconfigurable computing fabric has been proposed using different reconfigurable devices such as CNTFETs [Jab+11, CBO17] or SiNWFETs [Gai+15, DW04, Bob+12]. A summary of a reconfigurable fabric using emerging technology [Rai+21] shows multiple such architectures that can benefit from transistor-level reconfigurability. This effort can be further supported with interesting circuit designs of components such as multiplexer [Rai+17] or crossbars.

2.3 Logic Synthesis Preliminaries

While a brief introduction about EDA and its various stages are given in Sect. 1.3, here are the preliminaries and terminologies relevant to understanding the logic synthesis and technology mapping contributions of the book are explained. It is expected that the reader is familiar with the basic concepts of Boolean algebra. For more details, the following references [Mic94, Sas12] can be referred to.

2.3.1 Circuit Model

A hardware description of a circuit is described using an abstraction, i.e., a representation that shows relevant features. A circuit is viewed through the lens of an abstraction [Mic94]. In the present thesis, we consider the *logic* abstraction to represent a given circuit, representing a *Boolean function* f defined by the rules of *Boolean algebra*.

2.3.2 Boolean Algebra

A Boolean algebra is defined by the set $\mathbb{B} \supseteq \mathbb{B} \equiv 0, 1$ and by three operations $\wedge$, $\vee$, $'$ representing the conjunction (AND), disjunction (OR), and complementation (INV). The standard Boolean algebra is a non-empty set of ($\mathbb{B}$, 0, 1, $\wedge$, $\vee$, $'$) (axiomatization offered by Huntington's postulates [Hun04]). A Boolean function is a mapping between Boolean spaces. An n-input, m-output completely specified function is a mapping denoted by $f\colon \mathbb{B}^n \rightarrow \mathbb{B}^m$.

2.3.3 Monotone Function and the Property of Unateness

A *monotone* or a *unate* function is a function that can be easily represented by AND or OR function only. A function $f(x_1, x_2, \ldots, x_n)$ is monotonically increasing if and only if

$$f(0, x_2, \ldots, x_n) \leq f(1, x_2, \ldots, x_n), \qquad \forall(x_2, \ldots, x_n) \tag{2.1}$$

The *monotonically increasing* function is also called as *positive unate* function. Conversely, if $f(0, x_2, \ldots, x_n) \geq f(1, x_2, \ldots, x_n), \forall(x_2, \ldots, x_n)$, then the function f is called as *monotonically decreasing* or *negative unate* function. In a Boolean expression written either as a Disjunctive Normal Form (DNF) or Conjunctive Normal Form (CNF), the function is said to be *unate* in a variable where the literal corresponding to that variable is present only in one phase. Similarly, if the literal is present in both the phase, the function is said to be *binate* in that variable. For example, a simple logic function AND of two variables (let us say a and b), (the logic function of $f = a \vee b$) is unate in both the variables. However, if we have the logic function XOR, represented as $f = a \oplus b$, then the function is said to be binate in both the variables.

Understanding unateness is important within the context of emerging reconfigurable nanotechnology. Implementation of negative unate function is straightforward for CMOS technology where there exists a separate pull-up and pull-down network. However, implementation of XOR in CMOS logic requires more number of transistors. In contrast, emerging reconfigurable nanotechnology, by the virtue of its device-level reconfigurability, can implement a binate function with less number of transistors [Ama+15]. By using the PG, a binate function can be implemented with less number of transistors as will be shown in Fig. 3.1.

An EDA flow for RFETs-based circuits starts with a given circuit model and then applies various algorithms to get the final physical layout of the starting circuit. As discussed in Sect. 1.3, a Boolean network undergoes three main stages—logic synthesis, technology-independent mapping, and physical synthesis. Various notations and definitions required for these three steps are explained as follows:

2.3.4 Logic Representations

Various data representations are used in logic synthesis and technology mapping algorithms. Primarily two main representations are used for logic synthesis.

Two-Level Representations

Logic functions can be represented in DNF or CNF form. They are also referred to as Sum-of-Products (SOP) (*maxterms* of function f) and Product-of-Sums (POS)(minterms of function f), respectively. A logic function in DNF is represented as a product of literals[2]

[2] A literal is either the normal or the complement form of a given Boolean variable in a Boolean space $\mathbb{B}$.

$$f = t_1 \vee t_2 \vee \ldots \vee t_n \tag{2.2}$$

where each term t_i is a product of literals. In case of CNF representations, the $\vee$ is replaced by $\wedge$. Two-level representations are compact and are useful for representing small functions with few variables.

Multi-Level Representations

A multi-level graph network (also referred to as a *logic network*) is a Direct Acyclic Graph (DAG) with nodes corresponding to logic gates and directed edges corresponding to wires connecting the gates. The edges are directed from the inputs to the outputs of the logic network. A logic network is *homogeneous* if each node of the graph represents the same logic function. The edges can either be normal or complemented. In a logic graph, for intermediate nodes, the number of incoming edges defines the fan-in of that node. Similarly, the number of outgoing edges defines the fan-out of that node. The terms logic network, Boolean network, logic graph, and circuit are used interchangeably in this thesis.

In contemporary logic synthesis and technology-independent mapping, primarily multi-level homogeneous networks are used for carrying out logic optimization and mapping stages. It uses a combination of both *Boolean* and *algebraic methods* to optimize a given logic network. Binary Decision Diagrams (BDDs) [Bry86] and And-Inverter Graphs (AIGs) [BM10] are the two popular homogeneous networks where the nodes are *2to1 multiplexer* and *2-inputs AND* logic, respectively. While the BDD representation is *canonical* [BRB90], the AIG representation is not canonical. A representation is canonical if for a particular Boolean function f, the representation is unique. Canonicity is an important property to have a single functionally equivalent node for a given Boolean function. Using techniques such as *functional reduction* leads to *semi-canonical* AIGs [MCB07] so that no two nodes in an AIG should have the same functionality but the same function can have two different structural representations.

Recently, a homogeneous network, Majority-Inverter Graph (MIG), has been proposed where each node implements the logic function of MAJORITY [AGD14, AGM16]. A MAJORITY logic function is defined as follows:

$$f = \langle x, y, z \rangle = (x \vee y) \wedge (y \vee z) \wedge (z \vee x) \tag{2.3}$$

MIG has a sound and complete axiomitization in Boolean space [Ama+16]. An MIG node can be transformed to an AND or OR node by biasing the third input to 0 or 1, respectively. As a consequence, MIGs provide better logic compaction as compared to AIGs [AGM16].

Various mixed-functionality logic representations such as Xor-Inverter Graphs (XAGs) [HFS17] or Xor-Majority Graphs (XMGs) [Haa+17, Chu+19] have been demonstrated in contrast to otherwise popular primitives such as AIG or MIG [AGM16]. These newer primitives offer more compact logic representations which enable better runtimes for logic optimization and minimization flows [Haa+17]. XMGs or XAGs offer efficient representations of parity functions, as the size of AIGs and MIGs does not scale well with higher n-input parity functions [Chu+19].

With the advent of newer nanotechnology, there is a greater need for developing more efficient logic and physical synthesis tools for these newer reconfigurable technologies.

References

[AGD14] Luca Amarú, Pierre-Emmanuel Gaillardon, and Giovanni De Micheli. "Majority-Inverter Graph: A Novel Data-Structure and Algorithms for Efficient Logic Optimization". In: *Proceedings of the 51st Annual Design Automation Conference*. DAC '14. San Francisco, CA, USA: ACM, 2014, 194:1–194:6. https://doi.org/10.1145/2593069.2593158.

[AGM16] L. Amarú, P. E. Gaillardon, and G. De Micheli. "Majority-Inverter Graph: A New Paradigm for Logic Optimization". In: *IEEE Transactions on Computer-Aided Design of Integrated Circuits and Systems* 35.5 (May 2016), pp. 806–819. https://doi.org/10.1109/TCAD.2015.2488484.

[Ama+15] L. Amarú, P. E. Gaillardon, S. Mitra, and G. De Micheli. "New Logic Synthesis as Nanotechnology Enabler". In: *Proceedings of the IEEE* 103.11 (Nov. 2015), pp. 2168–2195. https://doi.org/10.1109/JPROC.2015.2460377.

[Ama+16] L. Amarú, P. E. Gaillardon, A. Chattopadhyay, and G. De Micheli. "A Sound and Complete Axiomatization of Majority-n Logic". In: *IEEE Transactions on Computers* 65.9 (Sept. 2016), pp. 2889–2895. https://doi.org/10.1109/TC.2015.2506566.

[Bi+16] Yu Bi, Kaveh Shamsi, Jiann-Shiun Yuan, Pierre-Emmanuel Gaillardon, Giovanni De Micheli, Xunzhao Yin, X Sharon Hu, Michael Niemier, and Yier Jin. "Emerging technology-based design of primitives for hardware security". In: *ACM Journal on Emerging Technologies in Computing Systems (JETC)* 13.1 (2016), pp. 1–19.

[BM10] Robert Brayton and Alan Mishchenko. "ABC: An Academic Industrial-Strength Verification Tool". In: ed. by Tayssir Touili, Byron Cook, and Paul Jackson. Berlin, Heidelberg: Springer Berlin Heidelberg, 2010, pp. 24–40. https://doi.org/10.1007/978-3-642-14295-6_5.

[Bob+12] Shashikanth Bobba, Michele De Marchi, Yusuf Leblebici, and Giovanni De Micheli. "Physical Synthesis Onto a Sea-of-Tiles with Double-gate Silicon Nanowire Transistors". In: *Proceedings of the 49th Annual Design Automation Conference*. DAC '12. San Francisco, California: ACM, 2012, pp. 42–47. https://doi.org/10.1145/2228360.2228369.

[BRB90] K.S. Brace, R.L. Rudell, and R.E. Bryant. "Efficient implementation of a BDD package". In: *27th ACM/IEEE Design Automation Conference*. 1990, pp. 40–45. https://doi.org/10.1109/DAC.1990.114826.

[Bry86] R. E. Bryant. "Graph-Based Algorithms for Boolean Function Manipulation". In: *IEEE Transactions on Computers* C-35.8 (Aug. 1986), pp. 677–691. https://doi.org/10.1109/TC.1986.1676819.

[CBO17] K. Cheng, S. Le Beux, and I. O'Connor. "Hybrid Topologies for Reconfigurable Matrices Based on Nano-Grain Cells". In: *2017 IEEE International Conference on Rebooting Computing (ICRC)*. Nov. 2017, pp. 1–8. https://doi.org/10.1109/ICRC.2017.8123639.

[Che+16] A. Chen, X. S. Hu, Y. Jin, M. Niemier, and X. Yin. "Using emerging technologies for hardware security beyond PUFs". In: *2016 Design, Automation Test in Europe Conference Exhibition (DATE)*. Mar. 2016, pp. 1544–1549.

[Chu+18] Chun-Lin Chu, Kehuey Wu, Guang-Li Luo, Bo-Yuan Chen, Shih-Hong Chen,Wen-FaWu, andWen-Kuan Yeh. "Stacked Ge-nanosheet GAAFETs fabricated by Ge/Si multilayer epitaxy". In: *IEEE Electron Device Letters* 39.8 (2018), pp. 1133–1136.

[Chu+19] Zhufei Chu, Mathias Soeken, Yinshui Xia, Lunyao Wang, and Giovanni De Micheli. "Structural Rewriting in XOR-Majority Graphs". In: *Proceedings of the 24th Asia and South Pacific Design Automation Conference*. ASPDAC '19. Tokyo, Japan: Association for Computing Machinery, 2019, pp. 663–668. https://doi.org/10.1145/3287624.3287671.

[Cou16] Rachel Courtland. "The next high-performance transistor [News]". In: *IEEE Spectrum* 53.10 (2016), pp. 11–12.

[DW04] Andre DeHon and Michael J. Wilson. "Nanowire-based Sublithographic Programmable Logic Arrays". In: *Proceedings of the 2004 ACM/SIGDA 12th International Symposium on Field Programmable Gate Arrays*. FPGA '04. Monterey, California, USA: ACM, 2004, pp. 123–132. https://doi.org/10.1145/968280.968299.

[Gai+13a] Pierre-Emmanuel Gaillardon, Luca Gaetano Amarù, Shashikanth Bobba, Michele De Marchi, Davide Sacchetto, Yusuf Leblebici, and Giovanni De Micheli. "Vertically-stacked Double-gate Nanowire FETs with Controllable Polarity: From Devices to Regular ASICs". In: *Proceedings of the Conference on Design, Automation and Test in Europe*. DATE '13. Grenoble, France: EDA Consortium, 2013, pp. 625–630.

[Gai+13b] Pierre-Emmanuel Gaillardon, Michele De Marchi, Luca Amarù, Shashikanth Bobba, Davide Sacchetto, Yusuf Leblebici, and Giovanni De Micheli. "Towards structured ASICs using polarity-tunable Si nanowire transistors". In: *Proceedings of the 50th Annual Design Automation Conference on - DAC '13* (2013), p. 1. https://doi.org/10.1145/2463209.2488886.

[Gai+15] Pierre Emmanuel Gaillardon, Xifan Tang, Gain Kim, and Giovanni De Micheli. "A Novel FPGA Architecture Based on Ultrafine Grain Reconfigurable Logic Cells". In: *IEEE Transactions on Very Large Scale Integration (VLSI) Systems* 23.10 (Oct. 2015), pp. 2187–2197. https://doi.org/10.1109/TVLSI.2014.2359385.

[Gor+19] G. Gore, P. Cadareanu, E. Giacomin, and P. Gaillardon. "A Predictive Process Design Kit for Three-Independent-Gate Field-Effect Transistors". In: *2019 IFIP/IEEE 27th International Conference on Very Large Scale Integration (VLSI-SoC)*. Oct. 2019, pp. 172–177. https://doi.org/10.1109/VLSI-SoC.2019.8920358.

[Haa+17] W. Haaswijk, M. Soeken, L. Amarù, P. E. Gaillardon, and G. De Micheli. "A novel basis for logic rewriting". In: *2017 22nd Asia and South Pacific Design Automation Conference (ASP-DAC)*. Jan. 2017, pp. 151–156. https://doi.org/10.1109/ASPDAC.2017.7858312.

[Har+10] Naoki Harada, Katsunori Yagi, Shintaro Sato, and Naoki Yokoyama. "A polarity-controllable graphene inverter". In: *Applied Physics Letters* (2010).

[Hei+12] André Heinzig, Stefan Slesazeck, Franz Kreupl, Thomas Mikolajick, and Walter M. Weber. "Reconfigurable silicon nanowire transistors". In: *Nano Letters* 12.1 (2012). PMID: 22111808, pp. 119–124. https://doi.org/10.1021/nl203094h. eprint: https://doi.org/10.1021/nl203094h.

[HFS17] I. H'alecek, P. Fišer, and J. Schmidt. "Are XORs in logic synthesis really necessary?" In: *2017 IEEE 20th International Symposium on Design and Diagnostics of Electronic Circuits Systems (DDECS)*. Apr. 2017, pp. 134–139. https://doi.org/10.1109/DDECS.2017.7934583.

[Hua+07] B. Huard, J. A. Sulpizio, N. Stander, K. Todd, B. Yang, and D. Goldhaber-Gordon. "Transport Measurements Across a TunablePotential Barrier in Graphene". In: *Phys. Rev. Lett.* 98 (23 June 2007), p. 236803. https://doi.org/10.1103/PhysRevLett.98.236803.

[Hun04] Edward V. Huntington. "Sets of Independent Postulates for the Algebra of Logic". In: *Transactions of the American Mathematical Society* 5.3 (1904), pp. 288–309.

[Jab+11] Kotb Jabeur, Natalya Yakymets, Ian O'Connor, and Sébastien Le-Beux. "Fine-grain reconfigurable logic cells based on double-gate CNTFETs". In: *Proceedings of the 21st edition of the great lakes symposium on Great lakes symposium on VLSI*. ACM. 2011, pp. 19–24.

[KM14] J. Knoch and M. R. Müller. "Electrostatic Doping—Controlling the Properties of Carbon-Based FETs With Gates". In: *IEEE Transactions on Nanotechnology 13.6* (2014), pp. 1044–1052. https://doi.org/10.1109/TNANO.2014.2323436.

[Lin+05] Y. Lin, J. Appenzeller, J. Knoch, and P. Avouris. "High-performance Carbon Nanotube Field-effect Transistor with Tunable Polarities". In: *TNano*. (2005).

[Lou+17] N Loubet, T Hook, P Montanini, C-W Yeung, S Kanakasabapathy, M Guillom, T Yamashita, J Zhang, X Miao, J Wang, et al. "Stacked nanosheet gate-all-around transistor to enable scaling beyond Fin-FET". In: *2017 Symposium on VLSI Technology*. IEEE. 2017, T230–T231.

[Mar+12] M. De Marchi, D. Sacchetto, S. Frache, J. Zhang, P. E. Gaillardon, Y. Leblebici, and G. De Micheli. "Polarity control in double-gate, gate-all-around vertically stacked silicon nanowire FETs". In: *2012 International Electron Devices Meeting*. Dec. 2012, pp. 8.4.1–8.4.4. https://doi.org/10.1109/IEDM.2012.6479004.

[Mar+14] M. De Marchi, D. Sacchetto, J. Zhang, S. Frache, P. E. Gaillardon, Y. Leblebici, and G. De Micheli. "Top-down fabrication of gateall-around vertically stacked silicon nanowire fets with controllable polarity". In: *IEEE Transactions on Nanotechnology* 13.6 (Nov. 2014), pp. 1029–1038. https://doi.org/10.1109/TNANO.2014.2363386.

[Mat+09] Takashi Matsukawa, Kazuhiko Endo, Yongxun Liu, Shinichi O'uchi, Yuki Ishikawa, Hiromi Yamauchi, Junichi Tsukada, Kenichi Ishii, Kunihiro Sakamoto, Eiichi Suzuki, and Meishoku Masahara. "Dual metal gate FinFET integration by Ta/Mo diffusion technology for Vt reduction and multi-Vt CMOS application". In: *Solid-State Electronics* 53.7 (2009). Papers Selected from the 38th European Solid-State Device Research Conference – ESSDERC'08, pp. 701–705. https://doi.org/10.1016/j.sse.2009.02.013.

[MCB07] A. Mishchenko, S. Chatterjee, and R. K. Brayton. "Improvements to Technology Mapping for LUT-Based FPGAs". In: *TCAD* 26.2 (2007), pp. 240–253.

[Mic94] Giovanni De Micheli. *Synthesis and optimization of digital circuits*. 1st. McGraw-Hill, 1994, p. 579.

[Mik+17] T Mikolajick, A Heinzig, J Trommer, T Baldauf, and W M Weber. "The RFET—a reconfigurable nanowire transistor and its application to novel electronic circuits and systems". In: *Semiconductor Science and Technology* 32.4 (2017), p. 043001.

[Mik+21] T Mikolajick, G Galderisi, M Simon, S Rai, A Kumar, A Heinzig, WM Weber, and J Trommer. "20 Years of reconfigurable field-effect transistors: From concepts to future applications". In: *Solid-State Electronics* 186 (2021), p. 108036.

[Mir+13] Sandeep Miryala, Mehrdad Montazeri, Andrea Calimera, Enrico Macii, and Massimo Poncino. "A Verilog-A model for reconfigurable logic gates based on graphene pn-junctions". In: *2013 Design, Automation Test in Europe Conference Exhibition (DATE)*. 2013, pp. 877–880. https://doi.org/10.7873/DATE.2013.185.

[MR16] Kirsten E Moselund and Heike E Riel. *Reconfigurable tunnel fieldeffect transistors*. US Patent 9,293,467. Mar. 2016.

[Nak+15] Shu Nakaharai, Mahito Yamamoto, Keiji Ueno, Yen-Fu Lin, Song-Lin Li, and Kazuhito Tsukagoshi. "Electrostatically reversible polarity of ambipolar a-MoTe2 transistors". In: *ACS Nano* 9.6 (2015). PMID: 25988597, pp. 5976–5983. https://doi.org/10.1021/acsnano.5b00736. eprint: https://doi.org/10.1021/acsnano.5b00736.

[Rai+17] M. Raitza, A. Kumar, M. Völp, D.Walter, J. Trommer, T. Mikolajick, and W. M. Weber. "Exploiting transistor-level reconfiguration to optimize combinational circuits". In: *Design, Automation Test in Europe Conference Exhibition (DATE), 2017*. Mar. 2017, pp. 338–343. https://doi.org/10.23919/DATE.2017.7927013.

[Rai+18a] S. Rai, A. Rupani, D. Walter, M. Raitza, A. Heinzig, T. Baldauf, J. Trommer, C. Mayr, W. M. Weber, and A. Kumar. "A physical synthesis flow for early technology evaluation of silicon nanowire based reconfigurable FETs". In: *2018 Design, Automation Test in Europe Conference Exhibition (DATE)*. Mar. 2018, pp. 605–608. https://doi.org/10.23919/DATE.2018.8342080.

[Rai+18b] Shubham Rai, Srivatsa Srinivasa, Patsy Cadareanu, Xunzhao Yin, Xiaobo Sharon Hu, Pierre-Emmanuel Gaillardon, Vijaykrishnan Narayanan, and Akash Kumar. "Emerging Reconfigurable Nanotechnologies: Can They Support Future Electronics?" In: *Proceedings of the International Conference on Computer-Aided Design*. ICCAD '18. San Diego, California: ACM, 2018, 13:1–13:8. https://doi.org/10.1145/3240765.3243472.

[Rai+19] S. Rai, J. Trommer, M. Raitza, T. Mikolajick, W. M. Weber, and A. Kumar. "Designing Efficient Circuits Based on Runtime-Reconfigurable Field-Effect Transistors". In: *IEEE Transactions on Very Large Scale Integration (VLSI) Systems* 27.3 (Mar. 2019), pp. 560–572. https://doi.org/10.1109/TVLSI.2018.2884646.

[Rai+20] Shubham Rai, Satwik Patnaik, Ansh Rupani, Johann Knechtel, Ozgur Sinanoglu, and Akash Kumar. "Security Promises and Vulnerabilities in Emerging Reconfigurable Nanotechnology-Based Circuits". In: *IEEE Transactions on Emerging Topics in Computing* (2020), pp. 1–1. https://doi.org/10.1109/TETC.2020.3039375.

[Rai+21] Shubham Rai, Pallab Nath, Ansh Rupani, Santosh Kumar Vishvakarma, and Akash Kumar. "A Survey of FPGA Logic Cell Designs in the Light of Emerging Technologies". In: *IEEE Access* 9 (2021), pp. 91564–91574. https://doi.org/10.1109/ACCESS.2021.3092167.

[Raj+15] J. Rajendran, R. Karri, J. B. Wendt, M. Potkonjak, N. McDonald, G. S. Rose, and B. Wysocki. "Nano Meets Security: Exploring Nanoelectronic Devices for Security Applications". In: *Proceedings of the IEEE* 103.5 (May 2015), pp. 829–849. https://doi.org/10.1109/JPROC.2014.2387353.

[Res+16] Giovanni V. Resta, Surajit Sutar, Yashwanth Balaji, Dennis Lin, Praveen Raghavan, Iuliana Radu, Francky Catthoor, Aaron Thean, Pierre-Emmanuel Gaillardon, and Giovanni de Micheli. "Polarity control in WSe2 double-gate transistors". In: *Scientific Reports* (2016).

[Reu+21] Maximilian Reuter, Johannes Pfau, Tillmann A. Krauss, Jürgen Becker, and Klaus Hofmann. "From MOSFETs to Ambipolar Transistors: Standard Cell Synthesis for the Planar RFET Technology". In: *IEEE Transactions on Circuits and Systems I: Regular Papers* 68.1 (2021), pp. 114–125. https://doi.org/10.1109/TCSI.2020.3035889.

[RRK19] A. Rupani, S. Rai, and A. Kumar. "Exploiting Emerging Reconfigurable Technologies for Secure Devices". In: *Euromicro DSD*. 2019.

[Sas12] T. Sasao. *Switching theory for logic synthesis*. Springer Science & Business Media, 2012.

[Sim+17] M. Simon, A. Heinzig, J. Trommer, T. Baldauf, T. Mikolajick, and W. M. Weber. "Top-Down Technology for Reconfigurable Nanowire FETs With Symmetric On-Currents". In: *IEEE Transactions on Nanotechnology* 16.5 (Sept. 2017), pp. 812–819. https://doi.org/10.1109/TNANO.2017.2694969.

[Sim+18] M. Simon, J. Trommer, B. Liang, D. Fischer, T. Baldauf, M. B. Khan, A. Heinzig, M. Knaut, Y. M. Georgiev, A. Erbe, J. W. Bartha, T. Mikolaiick, and W. M. Weber. "A wired-AND transistor: Polarity controllable FET with multiple inputs". In: *2018 76th Device Research Conference (DRC)*. June 2018, pp. 1–2. https://doi.org/10.1109/DRC.2018.8442159.

[Tan+10] S. Tanachutiwat, J. U. Lee, W. Wang, and C. Y. Sung. "Reconfigurable multi-function logic based on graphene p-n junctions". In: *Design Automation Conference*. June 2010, pp. 883–888. https://doi.org/10.1145/1837274.1837496.

[Ten+18] V. Tenace, A. Calimera, E. Macii, and M. Poncino. "Logic Synthesis of Pass-Gate Logic Circuits with Emerging Ambipolar Technologies". In: *IEEE Transactions on Computer-Aided Design of Integrated Circuits and Systems* (2018), pp. 1–1. https://doi.org/10.1109/TCAD.2018.2889770.

[Tro+14] Jens Trommer, André Heinzig, Anett Heinrich, Paul Jordan, Matthias Grube, Stefan Slesazeck, Thomas Mikolajick, and Walter M Weber. "Material Prospects of Reconfigurable Transistor (RFETs)–From Silicon to Germanium Nanowires". In: *MRS Online Proceedings Library Archive* 1659 (2014), pp. 225–230.

[Tro+15] Jens Trommer, André Heinzig, Tim Baldauf, Stefan Slesazeck, Thomas Mikolajick, and Walter M. Weber. "Functionality-Enhanced Logic Gate Design Enabled by Symmetrical Reconfigurable Silicon Nanowire Transistors". In: *IEEE Transactions on Nanotechnology* 14.4 (July 2015), pp. 689–698. https://doi.org/10.1109/TNANO.2015.2429893.

[Tro+16] J. Trommer, A. Heinzig, T. Baldauf, T. Mikolajick, W. M. Weber, M. Raitza, and M. Völp. "Reconfigurable nanowire transistors with multiple independent gates for efficient and programmable combinational circuits". In: *2016 Design, Automation Test in Europe Conference Exhibition (DATE)*. Mar. 2016, pp. 169–174.

[Tro+17a] J Trommer, A Heinzig, S Slesazeck, U Mühle, M Löffler, D Walter, C Mayr, T Mikolajick, and WM Weber. "Reconfigurable germanium transistors with low source-drain leakage for secure and energyefficient doping-free complementary circuits". In: *Device Research Conference (DRC), 2017 75th Annual*. IEEE. 2017, pp. 1–2.

[Tro+17b] Jens Trommer, André Heinzig, Uwe Mühle, Markus Löffler, Annett Winzer, Paul M. Jordan, Jürgen Beister, Tim Baldauf, Marion Geidel, Barbara Adolphi, Ehrenfried Zschech, Thomas Mikolajick, and Walter M. Weber. "Enabling Energy Efficiency and Polarity Control in Germanium Nanowire Transistors by Individually Gated Nanojunctions". In: *ACS Nano* 11.2 (2017). PMID: 28080025, pp. 1704–1711. https://doi.org/10.1021/acsnano.6b07531. eprint: https://doi.org/10.1021/acsnano.6b07531.

[Tro17] Jens Trommer. *Towards Reconfigurable Electronics by Functionality-Enhanced Circuits and Germanium Nanowire Devices*. BoD–Books on Demand, 2017.

[Tsc+02] J.W. Tschanz, J.T. Kao, S.G. Narendra, R. Nair, D.A. Antoniadis, A.P. Chandrakasan, and V. De. "Adaptive body bias for reducing impacts of die-to-die and within-die parameter variations on microprocessor frequency and leakage". In: *IEEE Journal of Solid-State Circuits* 37.11 (2002), pp. 1396–1402. https://doi.org/10.1109/JSSC.2002.803949.

[YEK19] Peide Ye, Thomas Ernst, and Mukesh V Khare. "The last silicon transistor: Nanosheet devices could be the final evolutionary step for Moore's Law". In: *IEEE Spectrum* 56.8 (2019), pp. 30–35.

[Yu+15] Lili Yu, Ahmad Zubair, Elton J. G. Santos, Xu Zhang, Yuxuan Lin, Yuhao Zhang, and Tom'as Palacios. "High-Performance WSe2 Complementary Metal Oxide Semiconductor Technology and Integrated Circuits". In: *Nano Letters* 15.8 (2015). PMID: 26192468, pp. 4928–4934. https://doi.org/10.1021/acs.nanolett.5b00668. eprint: https://doi.org/10.1021/acs.nanolett.5b00668.

[Zha+14a] J. Zhang, M. De Marchi, D. Sacchetto, P. E. Gaillardon, Y. Leblebici, and G. De Micheli. "Polarity-Controllable Silicon Nanowire Transistors With Dual Threshold Voltages". In: *IEEE Transactions on Electron Devices* 61.11 (Nov. 2014), pp. 3654–3660. https://doi.org/10.1109/TED.2014.2359112.

[Zha+14b] Jian Zhang, Xifan Tang, Pierre Emmanuel Gaillardon, and Giovanni De Micheli. "Configurable Circuits Featuring Dual-Threshold-Voltage Design With Three-Independent-Gate Silicon Nanowire FETs". In: *IEEE Transactions on Circuits and Systems I: Regular Papers* 61.10 (Oct. 2014), pp. 2851–2861. https://doi.org/10.1109/TCSI.2014.2333675.

Chapter 3
Exploring Circuit Design Topologies for RFETs

Designing logic gates based on emerging technologies is necessary as they form the building block for digital design. It lays the foundation to enable EDA for the commercialization of such a technology. With the emergence of reconfigurable nanotechnology as one of the viable technology for future electronics, a crucial task is to formalize designs and topologies for various logic gates and circuits [Rai+18b].

The unique property of ambipolarity in RFETs opens up new design paradigms for logic gates and circuits. The electrical symmetry offered by RFETs makes both p- and n-type functionality to be of equal drive strengths. This particular feature tends to avoid the transistor sizing issue with CMOS. Apart from ambipolarity, RFETs provide additional properties that should be utilized for building efficient circuits. First, there are certain types of RFETs (RFETs with nanowire geometry) with multiple gate terminals on a single channel to enable designs of logic gates with two or more inputs. As explained in the previous chapter, having multiple gate terminals on a single channel reduces the channel resistance. Reduction in resistance allows more inputs to be connected to a single transistor exhibiting the functionality of a wired-AND [Sim+18]. Hence, two or more input logic gates such as XOR3, MINORITY, or MAJORITY that were generally avoided in CMOS technology[1] can be realized with RFETs.

Consideration of the abovementioned transistor properties is integral for designing efficient circuits based on RFETs. Contemporary CMOS-styled design of logic gates can lead to sub-optimality in RFETs-based circuits in terms of physical properties such as area, delay, or power. Due to the above features, logic gates based on RFETs need to take special design consideration to efficiently utilize RFET's unique properties. Keeping this in mind, this chapter specifically targets the following research questions:

[1] While such logic gates can be realized in CMOS, due to cascading of multiple transistors between the output and V_{dd}, the larger logic gates are slower in CMOS implementation.

S. Rai, A. Kumar, *Design Automation and Applications for Emerging Reconfigurable Nanotechnologies*, https://doi.org/10.1007/978-3-031-37924-6_3

Research Questions Which circuit design topologies should be employed for logic gates and circuits based on RFETs? Can ambipolarity be abstracted for sequential or metastable circuits? This chapter introduces unique circuit designs based on RFETs and explores various opportunities and challenges for such circuits. It presents manually designed combinational and sequential circuit designs based on RFETs that can exploit the inherent ambipolarity.

3.1 Contributions

The major contributions of this chapter are as follows:

- A list of functionality-enhanced logic gates (the term *functionality-enhanced* was coined in [Tro+15]) based on RFETs is presented along with their detailed evaluation to facilitate design flow for RFET-based circuits.
- Functioning of an RFET-based metastable circuit using a *minority*-based SR latch that allows reconfiguration between a NAND- and NOR-based SR latch is demonstrated.
- Design of a reconfigurable dual edge-triggered D-flip-flop using RFETs based on True-Single Phase Clock (TSPC) logic is proposed which allows sampling at both the edges of the clock.
- Novel 1-bit Arithemetic Logic unit (ALU) design is implemented with RFETs and comparison to existing CMOS implementations has been shown to demonstrate how RFETs-centric considerations during design time lead to efficient circuits based on RFETs.

3.2 Organization

The present chapter is organized as follows: Sect. 3.3 presents the related work done in exploring circuits based on RFETs. Section 3.4 presents design topologies of exemplary seven reconfigurable logic gates, presenting their Boolean expression. Section 3.5 lists down logic gates that demonstrate structurally invariable logic for both values of program gate input. This is followed by Sect. 3.6 that shows an implementation of sequential circuits using RFETs. It also introduces how metastability can exploit ambipolarity for efficient circuit design based on RFETs. Section 3.7 presents a comparison of various circuit implementations based on RFETs and CMOS in terms of their area and delay calculation (based on logical effort). It discusses how novel design topology is needed for circuit design using an example of 1-bit ALU. Concluding remarks are given in Sect. 3.8.

3.3 Related Works

An important aspect is to use device-level (re)programmability at the circuit level to realize a runtime-reconfigurable circuit offering multiple functionalities. DeMarchi et al. have shown that XOR functionality is embedded naturally within dually gated reconfigurable devices [Mar+12]. Efficient arithmetic logic gates [AGM13, GAM14, OCo+07, OCo+12] and circuits [Gai+14, Gai+13a, Tur+13] based on RFETs have been demonstrated. Another important work, which first introduced simple logic gates based on RFETs was [Tro+15]. They evaluated various logic gate designs possible with logical effort theory. Raitza in [Res+17] and Gaillardon in [Gai+13b] showed good savings in area and delay for larger circuits and ASICs, respectively. By using carefully designed circuits for conditional carry adder, the authors showed savings in terms of number of transistors and delay in terms of logical effort [Res+17]. Quantitative analysis in terms of parameter numbers clearly reveals that conventional circuit designs are sub-optimal for newer nanotechnology and newer designs are essential for their true evaluation.

3.4 Exploring Design Topologies for Combinational Circuits: Functionality-Enhanced Logic Gates

Interesting combinational logic gates using RFETs can be realized. Typical CMOS-styled static logic gates are trivial to replicate in RFETs. All the PMOS transistors in the pull-up network need to be substituted by RFETs with program gate input connected to logic 0 (so that the RFET function as a p-type transistor). Similarly, all the NMOS transistors are to be replaced by RFETs in the n-type configuration.

3.4.1 List of Combinational Functionality-Enhanced Logic Gates Based on RFETs

However, ambipolarity enables designs of functionality-enhanced logic gates that can exhibit runtime- reconfigurability. In this section, seven such logic gate designs are proposed followed by a discussion on estimation of their gate delays using the logical effort theory.

The functionality-enhanced logic gates have been enumerated below:

1. **2-NAND-2NOR**: Fig. 3.1a shows a *MIN* logic gate that can be reconfigured to *2NAND* and *2NOR* functionality. The value of *P* determines the final functionality of the *MIN* logic gate. As shown in the figure, while logic 0 at the program gate input delivers two-input NAND, logic 1 at the program gate input delivers two-input NOR functionality. If a third input is connected to P, the

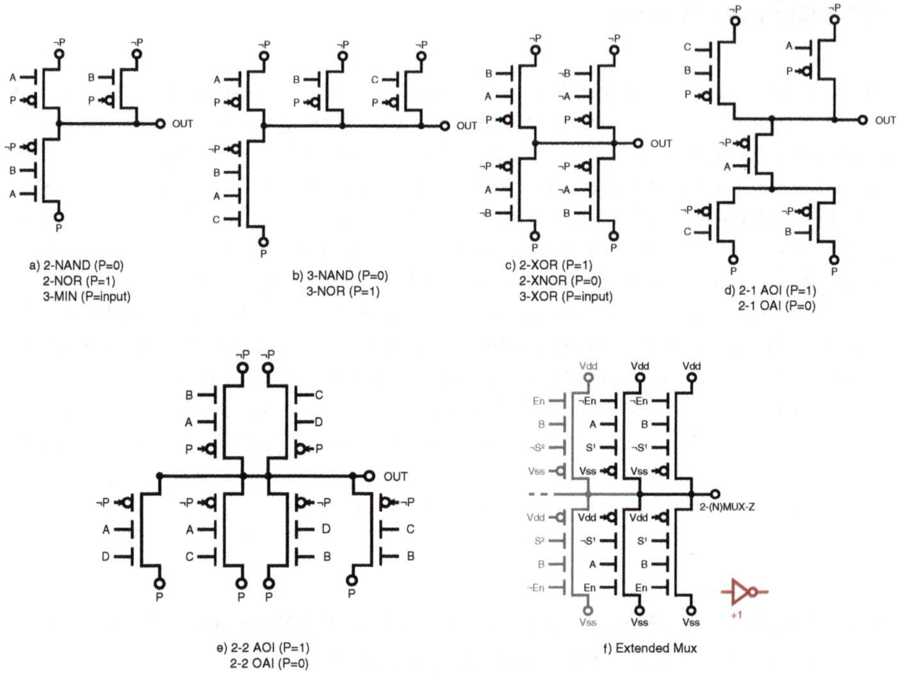

Fig. 3.1 Efficient combinational logic gates built from multi-independent-gate RFET technology: (**a**) 2NAND and 2NOR, (**b**) 3NAND and 3NOR, (**c**) 2XOR and 2XNOR, (**d**) 2-to-1 AOI and OAI, (**e**) 2-to-2 AOI and OAI, and (**f**) extended MUX functionality. In the static case, the program signal P is set to GND (0) or V_{DD} (1). A dynamic switching between both functions can be achieved by altering the program signal. In addition gates (**a**) and (**c**) can be executed in a transmission gate style by applying the program gate as additional input signal to map the 3MIN and 3XOR function, respectively

logic gate functions as a MIN gate. The technical implementation is referred to as the pass transistor logic [Tro+15], since an inverter or network passes the rail voltages to the functional cell. The Boolean function can be represented as:

$$f = P * (A + B)' + P' * (A * B)' \tag{3.1}$$

2. **2-AND-OR**: A trivial extension of the 2NAND-2NOR gate is achieved by adding an output inverter. Additionally, it can be written in POS form. The resulting 2AND-2OR Boolean function is represented by:

$$f = (P * (A + B)' + P' * (A * B)')' = (P' + (A + B)(P + (AB) \tag{3.2}$$

3. **3NAND-3NOR**: Fig. 3.1b shows a 3NAND-3NOR logic gate. The structure shows three RFETs connected in parallel with one multi-independent-gate (four-gate terminals) RFET connected at the bottom. Nanowire-based RFETs offer to

place multiple gates on a single wire without penalty in performance [Res+17, Sim+18]. The Boolean function is represented as:

$$f = P * (A + B + C)' + P' * (A * B * C)' \quad (3.3)$$

4. **2XOR-2XNOR:** Fig. 3.1c represents the *3-XOR* logic gate functionality. The potential at P determines the actual function of the logic gate. If P is fed with the third input, then the logic gate function changes from a two-input XOR/XNOR to a three-input XOR gate. The Boolean function is represented as:

$$f = P * (A * B + A' * B') + P' * (A * B' + A' * B) \quad (3.4)$$

5. **2-1 AOI-OAI:** Fig. 3.1d represents an *2-1 AOI-OAI*. This logic gate was first demonstrated in [Tro+15], but the representation was only based on dual-gate RFETs. Here we have used a combination of RFETs with a varying number of gate terminals on the channel to provide additional area savings. The Boolean function is represented as:

$$f = P * (A * B + C)' + P' * ((A + B) * C)' \quad (3.5)$$

6. **2-2 AOI-OAI:** Fig. 3.1e represents the *2-2 AOI-OAI* logic gate. The logic gate is similar to the 2-1 AOI-OAI version. The Boolean function is represented as:

$$f = P * (A * B + C * D)' + P' * ((A + B) * (C + D))' \quad (3.6)$$

7. **Extended MUX:** This logic gate implementation on RFETs uses multi-independent-gate terminals to encapsulate more logic. Figure 3.1g represents the extended multiplexer as used in [Res+17]. Raitza et al. used this *extended MUX* as a replacement for a two-stage multiplexer with an additional tristate-enabled signal E_n. The Boolean function is represented as:

$$f = E_n * (A * S_1' + B * (S_1 + S_2')) + E_n' * (A * S_1 + B * (S_1' + S_2)) \quad (3.7)$$

3.4.2 Estimation of Gate Delay Using the Logical Effort Theory

In order to make use of the above list of logic gates, it is important to find a measure for the performance of the above gates. Here, the delay of the proposed circuits is analyzed using the *logical effort theory* [SSH99] as described in [Tro+15]. The huge advantage of this method is that it delivers technology-independent results, which are directly transferable from a micrometer-sized lab technology to highly integrated circuits. By reformulation of a simple RC-based model, the propagation delay t_{PD} through an arbitrary logic gate can be described by:

$$t_{PD} = \tau * D \tag{3.8}$$

with

$$D = gh + p, \tag{3.9}$$

where τ is the intrinsic inverter delay, D is the structural delay of the circuit, h is the fanout, p is called parasitic delay, and g is called the logical effort, which is a direct measure for the logic inputs' topological complexity. A high logical effort is thereby associated with a high amount of input capacitance (which has to be charged), and thus, a larger circuit delay till the logic gate has successfully performed the output calculation.

However, for an overall delay comparison, the intrinsic inverter delay τ, which is a measure for the performance of the integrated technology, also has to be accounted for. In a first-order approximation, the intrinsic delay is inversely proportional to the on-currents I_{ON} of the individual device:

$$\tau = V_{DD} * C_G / I_{ON} \tag{3.10}$$

where V_{DD} is the supply voltage and C_G the input capacitance of a single-input gate.

With fabricated demonstrator devices still lacking in terms of on-current, the performance calculations done here can be easily transferred to a future highly scaled integrated technology. Recent simulation studies have shown that scaling of device dimensions, as well as applying several stacked nanowires on top of each other [Mar+14, Rai+17], increases the device performance of RFETs significantly. Moreover, promising performance projections are given for both germanium nanowires and carbon nanotubes [Bla+17, Tro+17b, Tro17, Tro+17a, RG18] making it conceivable that similar delay values as state-of-the-art low-operation-power CMOS technology can be achieved. It should be noted that the real intrinsic delay value of an RFET technology is somewhere in the middle of an inverter circuit utilizing only low-leakage-type Schottky gates or high-performance channel gates (refer to Fig. 2.1a). This is especially important within the critical path of a circuit, where only fast inputs should be used. However, utilizing the multigate technology, one can always add more gate terminals to the channel and trade area for a better transient performance.

τ is treated as a technology-agnostic value for all analysis under the assumption that an intrinsic delay similar to that of a respective CMOS device can be achieved by implementing the measures described above. Under this assumption, logical effort and normalized delay can be used as a direct measure to compare the performance of a certain circuit layout. In the first analysis step, the logic gates shown in Fig. 3.1 are assumed to be operating in a static mode. This means that all instances of P are directly connected to GND and all instances of $\overline{P}$ are directly connected to the supply voltage delivering a fixed functionality. In this configuration, the logic gates match their CMOS counterparts in terms of functionality.

3.5 Invariable Design of Inverters

Due to the available transistor-level ambipolarity, RFETs allow designs of structurally "invariable" logic. The term "invariable" refers to structural or schematic invariance while designing logic gates. This further implies that the functionality of logic gates remains the same when there is a change in the program gate input terminal. Figure 3.2 shows an RFET-based inverter, as inspired from designs in [RRK18] and [Tro+15]. The design shown in Fig. 3.2a is a static design. The P and $\overline{P}$ terminals are pre-connected to V_{dd} and V_{ss}, respectively. Figure 3.2b shows an inverter that is connected in a way that this cell only functions as an inverter when the value of P is at logic 0. This is so because the bottom transistor is already configured to be an n-type FET. Therefore, for this logic cell to function as an inverter, the upper transistor has to be configured as a p-type FET which requires P to be at logic 0. Similarly, the inverter design shown in Fig. 3.2c functions as an inverter only if the program gate input to this design is at logic 1. Figure 3.2d shows the completely invariant RFET-based inverter. Its layout is shown in Fig. 3.2e. The layout shows two inverters where the left inverter is driving the program gate inputs

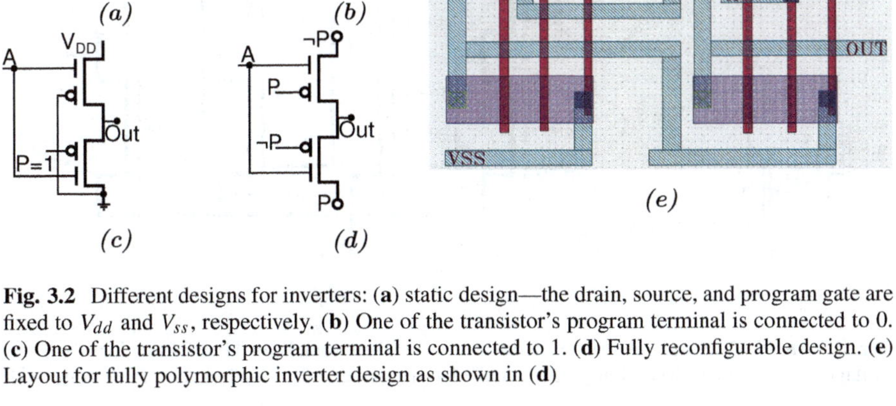

Fig. 3.2 Different designs for inverters: (**a**) static design—the drain, source, and program gate are fixed to V_{dd} and V_{ss}, respectively. (**b**) One of the transistor's program terminal is connected to 0. (**c**) One of the transistor's program terminal is connected to 1. (**d**) Fully reconfigurable design. (**e**) Layout for fully polymorphic inverter design as shown in (**d**)

for the right inverter. This design functions as an inverter irrespective of the potential at the program gate terminal. This happens because depending on the value of the P input, the pull-up and pull-down networks change in compliance with the power and ground terminals.

3.6 Sequential Circuits

The previous section covered designs of combinational logic gates. This section covers the sequential circuit design of an RFET-based flip-flop and metastable circuits.

3.6.1 Dual Edge-Triggered TSPC-Based D-Flip-Flop

The authors in [Tan+14] proposed a design of a single edge-triggered True-Single Phase Clock (TSPC)-based D-flip-flop using RFETs that has a reduced transistor count and area than its CMOS counterpart [YS89]. The design employed a dual-threshold voltage configuration of the TIGFETs as shown in Fig. 3.3a for true single-phase operation. In this configuration, input $G1$ has a LVT and input $G2$ has a HVT (corresponding to lower leakage current). Such freedom in the configuration of the threshold voltage improves the speed of the flip-flop by reducing the parasitics,

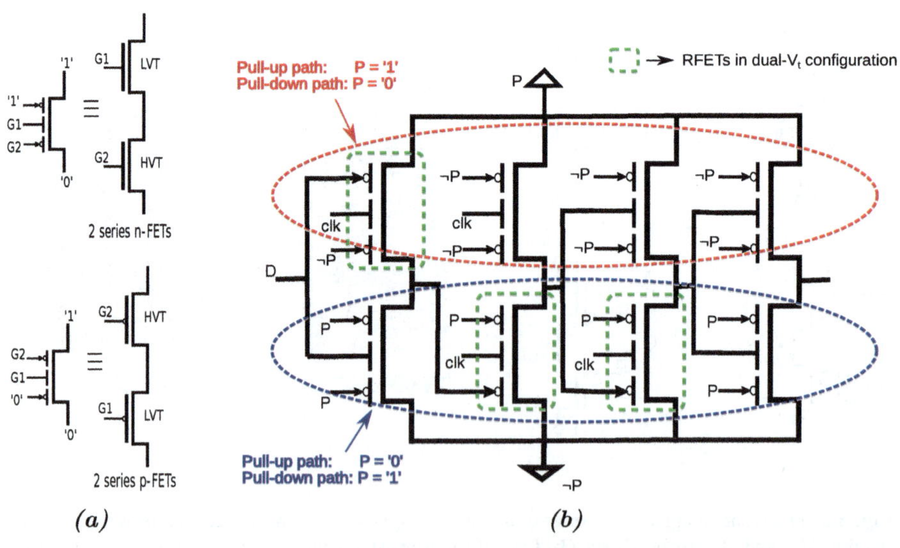

Fig. 3.3 (**a**) N-MOS and P-MOS transistor-level equivalent models for TIGFETs in dual-threshold voltage configuration. (**b**) A configurable dual edge-triggered D-flip-flop based on TSPC logic style

as each pull-up and pull-down path now consists of a single transistor [Tan+14, Zha+14]. It further leads to a reduced leakage power dissipation.

The runtime reconfigurability feature of RFETs is used to make the TSPC-based D-flip-flop proposed in [Tan+14] dual-edge triggered. This can be done by using a program signal (P) instead of the power rails as shown in Fig. 3.3b. If P = "1", the upper four transistors encircled in red provide the pull-up path, while the lower four transistors encircled in blue provide the pull-down path. In this case, the flip-flop samples data at the rising edge of the clock and hence behaves as a positive edge-triggered flip-flop. Conversely, if P = "0", the pull-up and pull-down paths get interchanged, and the flip-flop samples data at the falling clock edge. This way it behaves as a negative edge-triggered flip-flop.

3.6.2 *Exploiting RFET's Ambipolarity for Metastability*

In this section, the ambipolarity of individual transistors is used in a metastable circuit. Metastable circuits find huge applications in security as they are a crucial component while designing *physically unclonable functions* (PUFs) or *random number generators* (RNGs).

MINORITY Logic Gate-Based SR Latch

A metastable state can be attained by using cross-coupled elements as a source of randomness. Metastability-based circuits can be designed using reconfigurable *minority* (MIN) gate-based SR latch as shown in (Fig. 3.4a). Figure 3.4b shows a single SR latch unit consisting of two cross-coupled MIN gates and two buffers. Two clock signals ($clk\_Program$ and clk_IN) with the same time period T are

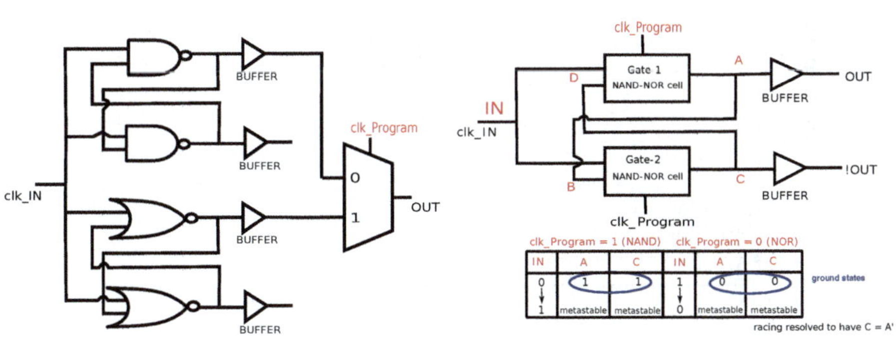

(a) (b)

Fig. 3.4 (**a**) A configurable *minority* (MIN) gate behaving as a NAND gate when P = "1" and NOR gate when P = "0". (**b**) An SR latch unit based on minority-based NAND-NOR cell fed into the unit, clk_IN being a time-delayed version of $clk\_Program$, delayed by t_d satisfying the condition $t_d < T/2$. In the first half-period of $clk\_Program$ ($clk\_Program$ = "1"), the MIN gates behave as NAND gates and as the rising edge of the clk_IN signal occurs; (when clk_IN = "0") the outputs of both the gates are "1" (ground state). After the transition in clk_IN signal, the outputs begin to race and temporarily enter into metastability. However, owing to the noise, the output "OUT" stabilizes in order to generate a random bit ("0" or "1"). Similarly, in the second half-period of $clk\_Program$ ($clk\_Program$ = "0"), the MIN gates behave as NOR gates and the falling edge of the clk_IN signal occurs. This time in the ground state, the outputs of both the gates are "0" and metastability is attained at the "1"→ "0" transition of the clk_IN signal, which eventually results in another random bit. Thus, in one complete clock cycle, two random bits are generated implying that the throughput of the SR latch unit is *twice* the input clock frequency.

The cross-coupled MIN gates (of same driving capability) in the SR latch unit (Fig. 3.4b) can be imagined as two cross-coupled inverters (such as in an SRAM cell) powered-ON, when the input clock makes a "0" → "1" transition for $clk\_Program$ = "1" or when it makes a "1" → "0" transition for $clk\_Program$ = "0". "*B*" and "*D*" are the inputs to *Gate-2* and *Gate-1*, respectively, while "*A*" and "*C*" are the outputs of *Gate-1* and *Gate-2*, respectively. The corresponding butterfly curve in the *voltage-transfer characteristic* (VTC) for the SR latch unit is shown in Fig. 3.5. This is done in Cadence Virtuoso where the simulation is carried for the circuit shown in Fig. 3.5b. It can be clearly seen that point "*X*", which is the point of metastability, lies on the identity line. This means that both stable states demarcated

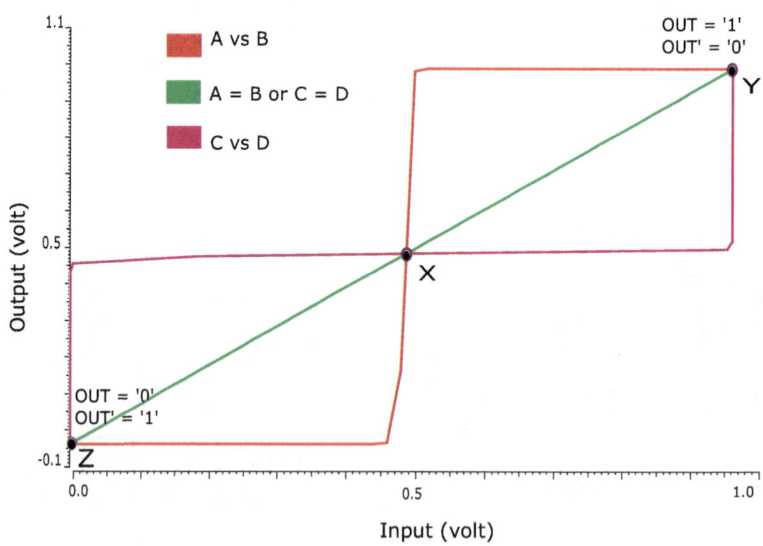

Fig. 3.5 Butterfly curve in the voltage-transfer characteristic (VTC) for the SR latch unit (Fig. 3.4b). "*B*" and "*D*" are the inputs to *Gate-2* and *Gate-1*, respectively, while "*A*" and "*C*" are the outputs of *Gate-1* and *Gate-2*, respectively

by points "Y" and "Z" are equally preferred. Eventually, the latch attains either state "Y" or "Z" due to noise, thereby producing a random bit at the output (OUT).

3.7 Evaluations

This section presents evaluation of logic gate designs shown till now.

3.7.1 Evaluation of Combinational Logic Gates

Calculation of normalized delay first requires values of g and p to be computed for the list of logic gates shown in Fig. 3.1.[2] Values of g and p for the respective logic gates are shown in Table 3.1. It is evident that due to their lower transistor count, all proposed multi-independent-gate RFET gates exhibit a reduced logical effort and less parasitic delay as compared to their CMOS counterparts. This performance increase is a result of the virtually lower channel resistance per input. In addition, there is always only a single transistor placed between the output node and the supply potentials. This omits the need for having several nanowires in parallel to speed up the serial branches. As a consequence, several changes in circuit topology become evident here. First of all, *NAND* and *NOR* circuits can be built with equal performance, simplifying timing constraints. Second, for the individual inputs all sorts of *NAND*, *NOR*, and *MUX* gates have a logical effort value *equal* to that of an inverter. This demonstrates that they all have equal driving strength. As a result, multigate RFETs provide an increased design flexibility as, e.g., all of those gates can be used to buffer a subsequent transmission gate, without an additional delay penalty. Moreover, it is evident that especially functions with a high number of inputs, such as *AOI* or *EMUX*, perform much better, when built using an RFET technology.

Overhead for Runtime Reconfiguration

If the proposed logic gates are applied for actual runtime reconfiguration, the program signals P and $\overline{P}$ have to be switched dynamically. In the most simple implementation, this can be executed by a single inverter routing the program signals [Tro+14]. For the sake of delay analysis, the whole gate including this input inverter can be described as a new form of the transmission gate. Logical effort for this type of gate is calculated using the methodology introduced in Trommer et al. [Tro+15]. The resulting logical effort and delay values are given in Table 3.2.

[2] Done in collaboration with Michael Raitza of the group.

Table 3.1 Comparison of transistor count #T, total logical effort gTot, logical effort per input signal gS, and parasitic delay p for static logic implementations of the gates depicted in Fig. 3.1

Gate	Variable	CMOS	MIGFET
Inverter	#T	2	2
	gTot	1	1
	gIN	1	1
	p	1	1
2-NAND	#T	4	3
	gTot	8/3	2
	gA/B	4/3	1
	p	2	3/2
2-NOR	#T	4	3
	gTot	10/3	2
	gA/B	5/3	1
	p	2	3/2
2-X(N)OR	#T	8	4
	gTot	8	4
	gA*/B*	4	2
	p	4	2
3-NAND	#T	6	4
	gTot	5	3
	gA/B/C	5/3	1
	p	3	2
3-NOR	#T	6	4
	gTot	7	3
	gA/B/C	7/3	1
	p	3	2
2-1AOI	#T	6	5
	gTot	17/3	9/2
	gA	5/3	3/2
	gB,C	2	3/2
	p	10/3	2
2-1OAI	#T	6	5
	gTot	16/3	9/2
	gA	4/3	3/2
	gB,C	2	3/2
	p	3	2
2-2 AOI	#T	8	6
	gTot	24	6
	gA,B,C,D	6	3/2
	p	6	2
2-MUX	#T	10	4
	gTot	8	4
	gA*,B*,C*	2	1
	p	4	2
2-EMUX	#T		6
	gTot		6
	gA*,B*,S*,En	N/A	1
	gS*,En		2
	p		2

Table 3.2 Comparison of transistor count #T, total logical effort gTot, logical effort per input signal gS and parasitic delay p for the 6 functionality-enhanced logic gates depicted in Fig. 3.1

Gate	Variable	MIGFET
2-NAND/2-NOR	#T	5
respective	gTot	11
3 min transmission	gA/B	2
	gp	7
	p	3
2-AND/OR (Fig. 3.1a + inv)	#T	7
	gTot	11
	gA/B	2
	gp	7
	p	4
2-XOR/2-XNOR	#T	6
respective	gTot	16
3-XOR Transmission	gA*/B*	4
	gp	8
	p	4
2-1AOI / 2-1 OAI	#T	7
	gTot	21
	gA/B/C	3
	gp	12
	p	4
2-2 AOI / 2-2 OAI	#T	8
	gTot	22
	gA/B/C/D	3
	gp	10
	p	6
	#T	6
3-NAND / 3-NOR	gTot	14
	gA/B/C	2
	gp	8
	p	4

It is obvious that the program signal comprises the largest logical effort (gP is much larger than gA/B/C) among all input signals. It implies that the process of reconfiguration takes more time than the actual processing of the input signals A and B. Further, as a trade-off for the increased functionality, the logical effort values for signals A and B are double as compared to the static logic gate implementation, due to the fact that there is an additional transistor placed between the supply potential and the output node.

Interestingly, some of the proposed gates enable additional functions when used as *pass-transistor logic* [Tro+15], which means that a logic input is applied at the outer source and drain contacts of the logic gate. As a result, for example, the 2NAND/2NOR or the 2XOR/XNOR gate inherently supports the 3MIN and 3XOR function, respectively, if P is used as the third logical input signal. However, it can be noted that reconfigurability is available at the cost of an increased number of transistors and logical effort. This is because of the slow reconfiguration paths, comprising of large capacitances which have to be addressed during operation. Thus, it is important to have a closer look at individual circuit designs, to exploit this added functionality. Next, an RFET-based ALU design is shown to exemplify these findings.

3.7.2 Novel Design of 1-Bit ALU

The authors in [RRK18] showed that ALU is also an example circuit that uses runtime reconfigurability. The reconfigurable nature of an ALU is possible due to the presence of 4-to-1 MUX which selects a specific functional output depending on the assignment of the select lines. In this case, the circuit computes all the functional outputs like AND, OR, XOR, and the full-adder. This is shown in Fig. 3.6a. It is to be noted that the ALU used here is a representative figure containing major components.

An innovative design of the representative 1-bit ALU using reconfigurable FET-based logic gates is shown in Fig. 3.6b. A comparative study with a CMOS-based circuit in terms of delay and area is carried out for the RFET-based circuit.

For the representative CMOS circuit (as shown in Fig. 3.6a), one can see that the AND, OR, and the NOT logic gate can be replaced by a single MIN gate (an additional inverter for AND and OR operation). With that approach, the AND-OR functionality of the ALU is replaced using a 3-RFET MIN logic gate and an inverter. This logic gate is used to calculate the *Cout* with *Cin* present. Hence, with a single *MIN* logic gate, one can generate all of the above functionality. For the full-adder implementation, the full-adder circuit in the normal ALU circuit can be removed and

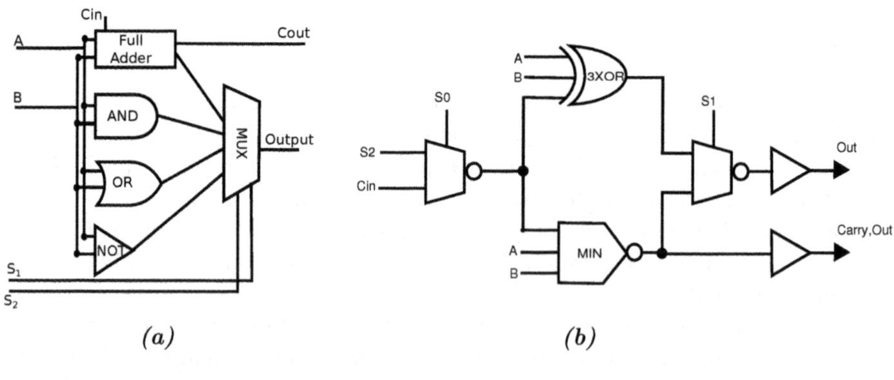

Fig. 3.6 (**a**) Gate-level representation of generic 1-bit ALU based on CMOS. (**b**) Novel gate-level representation of generic 1-bit ALU based on RFETs

replaced with a 3-bit XOR based on RFETs. Hence, the basic set of logic functions offered by the circuit of Fig. 3.6a can be realized using only these two logic gates. This circuit is referred to as the *ALU_reconf* in our calculations.

After this, a MUX is used to carefully connect the above two logic gates to select the required functionality. By intelligent use of S0 and S2, one can toggle between the various logic functionalities. The selection of S0, S1, and S2 to achieve different functionalities is shown in Table 3.3. The MUX with select lines S1 selects the output from the inputs from either the MIN or the 3-bit XOR. The first MUX selects the AND/OR functionality from the MIN gate to calculate *Cout*. The value of the select signal S0 needs to be 1 to select the *Cin* and to calculate the *Cout*. In other cases, where S0 is zero, S2 is passed which enables the circuit to have either the AND gate or the OR gate. For *Cout* calculation, MAJ logic function is used and *Cin* is the third input as evident from Table 3.3.

The truth table as shown in Table 3.3 also shows that this novel ALU can deliver other additional logic functions as well. The novel ALU design based on reconfigurable FETs is capable of producing more functions which is a bonus as compared to contemporary technologies. For example, *2-bit XNOR*, which is one of the most important functionality for equality comparison, can easily be achieved just by configuring the 3-bit XOR logic gate.

Results and Discussions

Table 3.4 shows the area, delay, and activity calculation for the existing and the novel 1-bit ALU circuit. Further, the original CMOS-based ALU design in terms of RFET (termed as *ALU_RFET* in Table 3.4) and included parameters for this version of circuit is considered as well. The CMOS circuit is the baseline reference for all the calculations.

For the actual area (in μm^2) comparison, the open-source library for CMOS at the 45-nm technology from *FreePDK45* [Sti+07] and the 22-nm SOI-based silicon nanowire RFET library [Rai+18a] with technology scaling [SXB] are used. Further, the area numbers for inverters in both the technologies are used to get an estimate of the post-physical synthesis area for the above circuits as shown in [Rai+18a]

Table 3.3 Novel ALU selection signals

S0	S1	S2	Out
0	0	0	NOR
0	0	1	NAND
0	1	0	XOR
0	1	1	XNOR
1	0	n/a	MAJ
1	0	n/a	MAJ
1	1	n/a	FA
1	1	n/a	FA

Table 3.4 Comparison in terms of area, normalized delay, activity, and the 1-bit ALU designs as shown in Fig. 3.6a and b

Circuit	Area (μm^2)	% area gain w.r.t. CMOS	Total normalized delay	% delay gain w.r.t. CMOS
ALU_CMOS (Fig. 3.6a)	5.54	Reference	36.47	Reference
ALU_RFETs (Fig. 3.6a)	9.50	−71.48	20.18	44.67
ALU reconf (Fig. 3.6b)	3.88	30.02	23.9	34.47

as all the logic gates are not available for RFETs. The area of inverter in RFET and CMOS technology (scaled to 22 nm) is *0.296* and *0.12* μm^2, respectively. To yield realistic area calculations, we use multiplication factors of 7/5 and 9/5 for three-independent-gate (TIG) RFETs [Tro+16] and 4-gated multi-independent-gate (MIG) RFETs [Tro+16] with respect to the simple dual-gate RFETs. The numbers are consistent with the finite-element models used to simulate the characteristics shown in Fig. 2.1a.

The area for novel ALU based on RFETs (*ALU_reconf*) is 30% smaller as compared to the CMOS counterpart. If the generic ALU (shown in Fig. 3.6a) is built from RFET, the area is 70% larger than that of the CMOS counterpart. That is coherent in terms of the sizes of individual transistors in RFET and CMOS technology and also backs our starting claims.

The nominal circuit delay of the circuits has been estimated using the method of logical effort. Thereto, the approximate minimal delay of each path through the circuits is calculated using the following expression:

$$D_{min} = N\big(h \prod g_i b_i\big)^{\frac{1}{N}} + \sum p_i \tag{3.11}$$

where N is the number of stages within the path; g_i and p_i are the logical effort and parasitic delay values of the individual stages, respectively; h is the fanout of the whole path; and b_i is the branching effort of every stage. Each input is thereby considered to be fed by an inverter. A general fanout of 4 is used for calculation. The delay of the slowest individual path, the so-called critical path, is then considered the nominal delay of the whole circuit. As stated earlier, all results are under the assumption that a similar individual device performance of RFETs and CMOS devices can be achieved, e.g., with the use of high-performance germanium nanowire channels or 2D materials such as WSe_2.

In terms of delay, as calculated using the logic effort theory, it is to be noted that the RFET-based circuits have higher performance as compared to the CMOS circuit primarily because of the reduced critical path of the overall circuit which can be attributed to the transistor-level reconfigurability. The *ALU_RFETs* is faster than the *ALU_reconf* as the latter uses pass-transistor logic. The pass-transistor logic is slower in performance but gains in higher functional expression. The RFET-based circuits are 44% and 34%, respectively, faster than the CMOS-based circuit.

It has to be taken into account that multi-independent-gate RFETs comprise the combination of fast inputs (low $|Vt|$) as well as a slow input (high $|Vt|$) (compared with the characteristics given in Fig. 2.1a as control gate terminal IV characteristics have a steeper slope [Res+17]). Hence, care has to be taken in circuit design that the output from the first *MUX* is not connected to the input in pass-transistor logic as that will add to circuit delay. Hence, a clever optimization is to apply this input to one of the middle control gate inputs of the transistor. In this case, one of the signals—A or B—has to be connected to the pass-transistor logic.

3.7.3 Comparison of the Sequential Circuit with an Equivalent CMOS-Based Design

The MIN-based SR latch is employed in a representative *random number generator* (RNG) design as shown in Fig. 3.7. The SR latch for both technologies are considered for comparison.

The simulation of the RFET-based SR latch has been carried out in Cadence Virtuoso. The Verilog-A model for the RFET in three-independent-gate configuration (TIGFET) from [Gor+19] has been used during the circuit-level simulations.

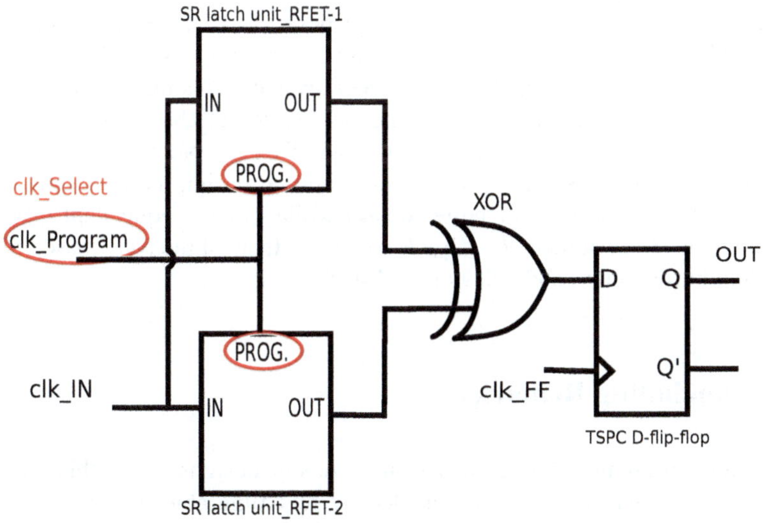

Fig. 3.7 A representative RNG schematic. The simulation model consisting of two SR latch units, an XOR gate in DG configuration and the configurable dual edge-triggered TSPC D-flip-flop based on RFETs. The CMOS implementation shown in Fig. 3.4a is used as the equivalent circuit for comparison

Table 3.5 A comparison between RFET-based SR latch unit and its CMOS equivalent

SR latch unit	Delay (ps)	μm^2
RFET	206	4.67
CMOS	909	3.9

This model has been adapted to incorporate flicker and white noise parameters. In the corresponding equivalent CMOS-based implementation (designed for double throughput) for the SR latch, operating at supply voltage of 1.0 V, the PTM model of 16-nm low-power CMOS model is used for the simulation of the MOSFETs [WY06]. The SR latch unit in this case is shown in Fig. 3.4a consisting of two cross-coupled NAND gates, two cross-coupled NOR gates, four buffers, and one 2×1 MUX.

In case of RFET-based design, the same circuit of the flip-flop can be reconfigured into both positive and negative edge-triggered functionalities based on the program signal during runtime. However, the same TSPC-based design of a D-flip-flop in CMOS technology [YS89] cannot be reconfigured as both positive and negative edge-triggered and it also uses more number of transistors (11 transistors) as compared to the proposed design using RFETs (8 transistors).

Table 3.5 presents a comparison between the SR latch units of the simulated proposed design and its CMOS counterpart (both for double throughput) on the basis of transistor count, power consumption, and delay operating at a clock frequency of 100 MHz. It can be seen that there is a 60% saving in the number of transistors by employing an RFET-based design. However, in terms of area calculation (in μm^2), RFET-based design has 19.91% overhead as compared to the CMOS design. Furthermore, a 94.5% reduction in leakage power, 70.7% reduction in dynamic power, and 77.3% reduction in path delay are observed in case of the SR latch unit based on RFETs with respect to its CMOS equivalent. This reduction in delay and hence power can be ascertained due to the fact that RFETs have lower parasitics as compared to the series connection of transistors [Tro+15]. Note, from Fig. 3.4b, that the path delay for the SR latch unit of the proposed design includes only clk_IN to "OUT" delay (inclusive of buffer delay) while, for the equivalent CMOS SR latch unit (Fig. 3.4a), it includes clk_IN to output delay of the NOR-based SR latch (including buffer delay) and delay of the MUX.

3.8 Concluding Remarks

This chapter presented new compact and efficient designs of combinational and sequential logic gates. These are enabled by reconfigurable nanowire transistors with multiple independent gates that can be used to replace arrangements of multiple transistors in series. This leads to several differences in circuit topology as compared to CMOS technology, e.g., NAND, NOR, and MUX all provide inverter drive capability. It has been found that the logic gates based on reconfigurable transistors are functionality-enhanced.

A novel design for a 1-bit ALU circuit based on RFET technology is presented. As compared to the CMOS technology, using this novel design achieves an area reduction of 30% and circuit delay reduction of 34%. It has been shown how efficient circuit design using reconfigurable transistors can lead to a range of benefits over contemporary CMOS technologies. The circuit shown in Fig. 3.6b gives a strong statement for the number of functions that can be achieved by using RFETs. Further, the functional range is higher as compared to the traditional ALU. Intelligent design approaches have to be taken in the case of circuits made of novel emerging nanotechnologies to truly harness their benefits.

The technology albeit in its infancy has shown a lot of promise in showing better numbers for area and delay. Various other works like [BM15, ZGD13, Gai+14], etc. have shown the efficacy of this technology in terms of area, power, and delay, respectively, catering to static logic. Note that the disruptive reconfigurable technology is not limited to silicon and is expandable to other semiconductor materials, like germanium [Tro17] and carbon [OCo+12, Bla+17]. The exquisite feature lies in the ease of extended functionality that RFET nanotechnology can provide.

References

[AGM13] L. Amarù, P. E. Gaillardon, and G. De Micheli. "Efficient arithmetic logic gates using double-gate silicon nanowire FETs". In: *2013 IEEE 11th International New Circuits and Systems Conference (NEWCAS)*. June 2013, pp. 1–4. https://doi.org/10.1109/NEWCAS.2013.6573572.

[Bla+17] S. Blawid, D. L. M. de Andrade, S. Mothes, and M. Claus. "Performance Projections for a Reconfigurable Tunnel NanoFET". In: *IEEE Journal of the Electron Devices Society* 5.6 (Nov. 2017), pp. 473–479. https://doi.org/10.1109/JEDS.2017.2756040.

[BM15] S. Bobba and G. De Micheli. "Layout Technique for Double-Gate Silicon Nanowire FETs With an Efficient Sea-of-Tiles Architecture". In: *IEEE Transactions on Very Large Scale Integration (VLSI) Systems* 23.10 (Oct. 2015), pp. 2103–2115. https://doi.org/10.1109/TVLSI.2014.2358884.

[Gai+13a] Pierre-Emmanuel Gaillardon, Luca Gaetano Amarù, Shashikanth Bobba, Michele De Marchi, Davide Sacchetto, Yusuf Leblebici, and Giovanni De Micheli. "Vertically-stacked Double-gate Nanowire FETs with Controllable Polarity: From Devices to Regular ASICs". In: *Proceedings of the Conference on Design, Automation and Test in Europe*. DATE '13. Grenoble, France: EDA Consortium, 2013, pp. 625–630.

[Gai+13b] Pierre-Emmanuel Gaillardon, Michele De Marchi, Luca Amarù, Shashikanth Bobba, Davide Sacchetto, Yusuf Leblebici, and Giovanni De Micheli. "Towards structured ASICs using polarity-tunable Si nanowire transistors". In: *Proceedings of the 50th Annual Design Automation Conference on - DAC '13* (2013), p. 1. https://doi.org/10.1145/2463209.2488886.

[Gai+14] Pierre Emmanuel Gaillardon, Shashikanth Bobba, Michele De Marchi, Davide Sacchetto, and Giovanni De Micheli. "Nanowire systems: technology and design". In: *Philosophical Transactions of the Royal Society of London* (2014).

[GAM14] P. E. Gaillardon, L. G. Amarù, and G. D. Micheli. "Unlocking Controllable-Polarity Transistors Opportunities by Exclusive-OR and Majority Logic Synthesis". In: *2014 IEEE Computer Society Annual Symposium on VLSI*. July 2014, pp. 403–405. https://doi.org/10.1109/ISVLSI.2014.107.

[Gor+19] G. Gore, P. Cadareanu, E. Giacomin, and P. Gaillardon. "A Predictive Process Design Kit for Three-Independent-Gate Field-Effect Transistors". In: *2019 IFIP/IEEE 27th International Conference on Very Large Scale Integration (VLSI-SoC)*. Oct. 2019, pp. 172–177. https://doi.org/10.1109/VLSI-SoC.2019.8920358.

[Mar+12] M. De Marchi, D. Sacchetto, S. Frache, J. Zhang, P. E. Gaillardon, Y. Leblebici, and G. De Micheli. "Polarity control in double-gate, gate-all-around vertically stacked silicon nanowire FETs". In: *2012 International Electron Devices Meeting*. Dec. 2012, pp. 8.4.1–8.4.4. https://doi.org/10.1109/IEDM.2012.6479004.

[Mar+14] M. De Marchi, D. Sacchetto, J. Zhang, S. Frache, P. E. Gaillardon, Y. Leblebici, and G. De Micheli. "Top-down fabrication of gateall- around vertically stacked silicon nanowire fets with controllable polarity". In: *IEEE Transactions on Nanotechnology* 13.6 (Nov. 2014), pp. 1029–1038. https://doi.org/10.1109/TNANO.2014.2363386.

[OCo+07] I. O'Connor, J. Liu, F. Gaffiot, F. Pregaldiny, C. Lallement, C. Maneux, J. Goguet, S. Fregonese, T. Zimmer, L. Anghel, T. Dang, and R. Leveugle. "CNTFET Modeling and Reconfigurable Logic- Circuit Design". In: *IEEE Transactions on Circuits and Systems I: Regular Papers* 54.11 (Nov. 2007), pp. 2365–2379. https://doi.org/10.1109/TCSI.2007.907835.

[OCo+12] Ian O'Connor, Kotb Jabeur, Sébastien Le Beux, and David Navarro. "Ambipolar Independent Double Gate FET Logic". In: *Proceedings of the 2012 IEEE/ACM International Symposium on Nanoscale Architectures*. NANOARCH '12. Amsterdam, The Netherlands: ACM, 2012, pp. 61–68. https://doi.org/10.1145/2765491.2765504.

[Rai+17] M. Raitza, A. Kumar, M. Völp, D.Walter, J. Trommer, T. Mikolajick, and W. M. Weber. "Exploiting transistor-level reconfiguration to optimize combinational circuits". In: *Design, Automation Test in Europe Conference Exhibition (DATE), 2017*. Mar. 2017, pp. 338–343. https://doi.org/10.23919/DATE.2017.7927013.

[Rai+18a] S. Rai, A. Rupani, D. Walter, M. Raitza, A. Heinzig, T. Baldauf, J. Trommer, C. Mayr, W. M. Weber, and A. Kumar. "A physical synthesis flow for early technology evaluation of silicon nanowire based reconfigurable FETs". In: *2018 Design, Automation Test in Europe Conference Exhibition (DATE)*. Mar. 2018, pp. 605–608. https://doi.org/10.23919/DATE.2018.8342080.

[Rai+18b] Shubham Rai, Srivatsa Srinivasa, Patsy Cadareanu, Xunzhao Yin, Xiaobo Sharon Hu, Pierre-Emmanuel Gaillardon, Vijaykrishnan Narayanan, and Akash Kumar. "Emerging Reconfigurable Nanotechnologies: Can They Support Future Electronics?" In: *Proceedings of the International Conference on Computer-Aided Design*. ICCAD '18. San Diego, California: ACM, 2018, 13:1–13:8. https://doi.org/10.1145/3240765.3243472.

[Res+17] Giovanni V Resta, Tarun Agarwal, Dennis Lin, Iuliana P Radu, Francky Catthoor, Pierre-Emmanuel Gaillardon, and Giovanni De Micheli. "Scaling trends and performance evaluation of 2-dimensional polarity-controllable FETs". In: *Scientific Reports* 7 (2017), p. 45556.

[RG18] J. Romero-González and P. Gaillardon. "BCB Evaluation of High- Performance and Low-Leakage Three-Independent-Gate Field-Effect Transistors". In: *IEEE Journal on Exploratory Solid-State Computational Devices and Circuits* 4.1 (June 2018), pp. 35–43. https://doi.org/10.1109/JXCDC.2018.2821638.

[RRK18] S. Rai, M. Raitza, and A. Kumar. "Technology mapping flow for emerging reconfigurable silicon nanowire transistors". In: *2018 Design, Automation Test in Europe Conference Exhibition (DATE)*. Mar. 2018, pp. 767–772. https://doi.org/10.23919/DATE.2018.8342110.

[Sim+18] M. Simon, J. Trommer, B. Liang, D. Fischer, T. Baldauf, M. B. Khan, A. Heinzig, M. Knaut, Y. M. Georgiev, A. Erbe, J. W. Bartha, T. Mikolaiick, and W. M. Weber. "A wired-AND transistor: Polarity controllable FET with multiple inputs". In: *2018 76th Device Research Conference (DRC)*. June 2018, pp. 1–2. https://doi.org/10.1109/DRC.2018.8442159.

[SSH99] Ivan Edward Sutherland, Robert F Sproull, and David F Harris. *Logical effort: designing fast CMOS circuits*. Morgan Kaufmann, 1999.

[Sti+07] J. E. Stine, I. Castellanos, M. Wood, J. Henson, F. Love, W. R. Davis, P. D. Franzon, M. Bucher, S. Basavarajaiah, J. Oh, and R. Jenkal. "FreePDK: An Open-Source Variation-Aware Design Kit". In: *2007 IEEE International Conference on Microelectronic Systems Education (MSE'07)*. June 2007, pp. 173–174. https://doi.org/10.1109/MSE.2007.44.

[SXB] Aaron Stillmaker, Zhibin Xiao, and Bevan Baas. "Toward more accurate scaling estimates of cmos circuits from 180 nm to 22 nm". In: *Technical Report ECE VCL 2011 4 VLSI Computation Lab, University of California,Davis* ().

[Tan+14] X. Tang, J. Zhang, P. Gaillardon, and G. De Micheli. "TSPC Flip- Flop circuit design with three-independent-gate silicon nanowire FETs". In: *2014 IEEE International Symposium on Circuits and Systems (ISCAS)*. June 2014, pp. 1660–1663. https://doi.org/10.1109/ISCAS.2014.6865471.

[Tro+14] J. Trommer, A. Heinzig, S. Slesazeck, T. Mikolajick, and W. M. Weber. "Elementary Aspects for Circuit Implementation of Reconfigurable Nanowire Transistors". In: *IEEE Electron Device Letters* 35.1 (Jan. 2014), pp. 141–143. https://doi.org/10.1109/LED.2013.2290555.

[Tro+15] Jens Trommer, André Heinzig, Tim Baldauf, Stefan Slesazeck, Thomas Mikolajick, and Walter M. Weber. "Functionality-Enhanced Logic Gate Design Enabled by Symmetrical Reconfigurable Silicon Nanowire Transistors". In: *IEEE Transactions on Nanotechnology* 14.4 (July 2015), pp. 689–698. https://doi.org/10.1109/TNANO.2015.2429893.

[Tro+16] J. Trommer, A. Heinzig, T. Baldauf, T. Mikolajick, W. M. Weber, M. Raitza, and M. Völp. "Reconfigurable nanowire transistors with multiple independent gates for efficient and programmable combinational circuits". In: *2016 Design, Automation Test in Europe Conference Exhibition (DATE)*. Mar. 2016, pp. 169–174.

[Tro+17a] J Trommer, A Heinzig, S Slesazeck, U Mühle, M Löffler, D Walter, C Mayr, T Mikolajick, and WM Weber. "Reconfigurable germanium transistors with low source-drain leakage for secure and energyefficient doping-free complementary circuits". In: *Device Research Conference (DRC), 2017 75th Annual*. IEEE. 2017, pp. 1–2.

[Tro+17b] Jens Trommer, André Heinzig, Uwe Mühle, Markus Löffler, Annett Winzer, Paul M. Jordan, Jürgen Beister, Tim Baldauf, Marion Geidel, Barbara Adolphi, Ehrenfried Zschech, Thomas Mikolajick, and Walter M. Weber. "Enabling Energy Efficiency and Polarity Control in Germanium Nanowire Transistors by Individually Gated Nanojunctions". In: *ACS Nano* 11.2 (2017). PMID: 28080025, pp. 1704–1711. https://doi.org/10.1021/acsnano.6b07531. eprint: http://doi.org/10.1021/acsnano.6b07531.

[Tro17] Jens Trommer. *Towards Reconfigurable Electronics by Functionality- Enhanced Circuits and Germanium Nanowire Devices*. BoD–Books on Demand, 2017.

[Tur+13] Ogun Turkyilmaz, Fabien Clermidy, Luca Gaetano Amaru, Pierre Emmanuel Gaillardon, and Giovanni De Micheli. "Self-checking ripple-carry adder with Ambipolar Silicon NanoWire FET". In: *Proceedings - IEEE International Symposium on Circuits and Systems* (May 2013), pp. 2127–2130. https://doi.org/10.1109/ISCAS.2013.6572294.

[WY06] Wei Zhao and Yu Cao. "New generation of predictive technology model for sub-45nm design exploration". In: *ISQED*. 2006.

[YS89] J. Yuan and C. Svensson. "High-speed CMOS circuit technique". In: *JSSC* (1989).

[ZGD13] Jian Zhang, Pierre Emmanuel Gaillardon, and Giovanni De Micheli. "Dual-threshold-voltage configurable circuits with three-independentgate silicon nanowire FETs". In: *Proceedings - IEEE International Symposium on Circuits and Systems*. May 2013, pp. 2111–2114. https://doi.org/10.1109/ISCAS.2013.6572291.

[Zha+14] Jian Zhang, Xifan Tang, Pierre Emmanuel Gaillardon, and Giovanni De Micheli. "Configurable Circuits Featuring Dual-Threshold-Voltage Design With Three-Independent-Gate Silicon Nanowire FETs". In: *IEEE Transactions on Circuits and Systems I: Regular Papers* 61.10 (Oct. 2014), pp. 2851–2861. https://doi.org/10.1109/TCSI.2014.2333675.

Chapter 4
Standard Cells and Technology Mapping

The previous chapter focused on various kinds of logic gates and circuit-design topologies based on reconfigurable emerging technologies. These logic gates and circuits helped us to understand the logical properties of RFETs in circuit design. However, to advance an emerging technology as a viable commercial technology, it is imperative to devise automated techniques to build circuits based on such technologies.

This chapter takes a step towards an automated EDA flow for emerging reconfigurable nanotechnology. While most of the prior works in the literature targeted the logic optimization and minimization of electronic circuits for RFET-based circuits [AGD14b, Ama+15, AGD13, AGD14a], this chapter proposes a technology mapping flow that utilizes the exceptional property of functional reconfigurability exhibited by RFET-based circuits. The technology mapping flow allows defining a range of functionality for a single logic gate to be used during mapping. This helps to provide an early evaluation for the circuit based on such multifunctional RFET logic gates.

Another interesting perspective that this chapter deals with is to understand the relation between a specific type of Boolean truth table and its RFET implementation. From the previous chapter, it was clear that while RFETs enable runtime reconfiguration between p-type and n-type behavior, not all logic gates are reconfigurable or exhibit multiple functionalities. There are logic gates that follow CMOS-style gate design and exhibit a conventional single functionality. Then, why do only particular types of logic gates based on RFETs demonstrate extended functionality? The chapter specifically targets the following research question:

Research Question Logic gates based on CMOS are governed by electrical characteristics of distinct pull-up and pull-down networks. Is the same CMOS-styled circuit design applicable for RFET-based logic gates? This chapter investigates the reason why only certain logic functions can be implemented as reconfigurable logic gates. Is there a universal Boolean property that can be implemented efficiently using RFETs?

S. Rai, A. Kumar, *Design Automation and Applications for Emerging Reconfigurable Nanotechnologies*, https://doi.org/10.1007/978-3-031-37924-6_4

This chapter demonstrates that logic gates based on self-dual functions are a preferred choice for standard-cell implementation based on RFETs. The chapter explores how the duality in the electrical property expressed due to ambipolarity at the transistor level transcends into duality at the logical level. In this regard, self-duality is shown to be the natural abstraction for RFET-based circuits. The present chapter also lays the foundation for the next chapter where logic representations are explored for efficient logic synthesis flows.

4.1 Contributions

The major contributions of this chapter are as follows:

1. A methodology is presented to identify Boolean logic that can exploit the intrinsic property of transistor-level duality to demonstrate reconfigurable functionality. These reconfigurable logic gates can be efficiently implemented using reconfigurable nanotechnology devices and can lead to optimized circuit implementation in terms of metrics like area, power, and delay.
2. An algorithm is proposed that can distill probable standard cells from a circuit's logic network based on truth tables of individual cuts of a mapped circuit. These probable standard cells are a direct outcome of the requirements of the logic implementation of individual circuits and hence lead to better mapping in terms of area, delay, and edge connections.
3. The concept of HOFs is introduced to encapsulate the functional reconfigurability offered by RFET-based logic gates. HOF-based mapping moves away from the normal CMOS flow in which logic gates have a single immutable function. This enables mapping one of the multiple yet mutually exclusive functions expressed by *HOFs* for reconfigurability-aware circuits.
4. Based on the transfer characteristics of RFETs, a heuristic to contribute to area savings is proposed for the HOF-based mapping. The mapping uses available inverted forms of fan-ins from the circuit for the family of XOR logic gates.

4.2 Organization

The present chapter is organized as follows: Sect. 4.3 presents the related works in the direction of technology-independent mapping for RFET-based circuits. Section 4.4 explains the relation between Boolean truth tables and logic gates that can be efficiently implemented using RFETs. It shows how self-duality forms the natural logical abstraction for RFET-based logic gates. Then Sect. 4.5 discusses the methodology to use the Boolean property of self-duality to find standard cells. The section describes an algorithm that is applied over a benchmark suite to extract self-dual standard cells. Section 4.6 introduces the mathematical foundation of

HOFs to abstract reconfigurable logic gates based on RFETs. The concept of HOFs is used to define a new style of RFET-based generic library. The concept of inverter adjustments for the XOR family of logic gates based on RFETs is also introduced in this section. This is followed by Sect. 4.7 that describes the experiments carried out in this chapter. Concluding remarks and future work directions are given in Sect. 4.8.

4.3 Related Work

In [Tro+16], Trommer et al. showed efficient and programmable combinational logic gates based on RFETs. It was shown that because of this extended functionality, a single transistor can replace logic by several transistors. Further, in [Rai+17], the authors demonstrated the potential of reconfigurable transistors in a conditional carry adder circuit by showing area and delay gains. These works showed the benefits of RFETs at the circuit level. MIG [AGM16] and MixSyn [AGD13] explored data structures and algorithms to exploit logic minimization and optimization flow. In [AGM13], Amaru et al. showed that majority functions are natural functional representation for SiNW FETs. In MixSyn, the authors tried to partition and employed a different optimization for XOR intrinsic parts of the circuits and *AND/OR* parts of the circuits. This tool explored the XOR implementations offered inherently by ambipolar devices as mentioned in [Mar+12]. None of these works truly exploited reconfigurability offered by RFETs at the technology mapping level and undermine their benefits by using an analogous CMOS flow. While [Tro+16] and [Rai+17] have demonstrated specific cases of individual gates to simple manual designed circuits, no automated mapping methodology has been proposed especially for RFETs.

In this work, a mapping flow is introduced for circuits based on RFETs by utilizing the flexibility in terms of functional output of RFET logic gates. The technology mapping flow helps to carry out early evaluation of the overheads when using the functionality-enhanced logic gates as proposed in the previous chapter. Further, a potential saving in the area by reusing inverter logic available in the circuit for XOR-based logic gates is also presented.

4.4 Standard Cells Based on RFETs

The present section explains the reason behind why only certain logic gates exploit the ambipolarity offered by individual RFETs to exhibit reconfigurability and not others.

4.4.1 Interchangeable Pull-Up and Pull-Down Networks

In conventional CMOS-based logic gates, both NMOS and PMOS devices are fundamentally and physically different from each other. This results in CMOS-based logic gates having a distinct Pull-up Network (PUN) made of PMOS transistors which drives the output of the logic gate to logic 1. The inverse is the Pull-down Network (PDN) consisting of NMOS transistors which drive the output to logic 0. This is shown in Fig. 4.1a. Since PMOS devices are slower as compared to NMOS (due to lesser mobility of holes as compared to electrons), designing logic gates based on CMOS often requires careful design of channel widths to achieve a similar drive strength of the Pull-up Network (PUN) and Pull-down Network (PDN).

In contrast, RFETs do not have such imbalance between the p and n-type behavior as they are functionally and electrically symmetric in both configurations [Hei+12]. Physically, the p- and n-type behavior are two different renditions of the same device. Hence, by changing the potential at the program gate (and correspondingly at the control gate so that the device is in the on-state), both p- and n-type behavior are realized from a single RFET. When this change of potential at the program gate is done to the logic gate, the pull-up and pull-down network can be interchanged, and hence, more than one logical functionality can be realized. In Fig. 4.1b, the potential change at *reconf_input* switches the PUN and PDN. For the list of logic gates described in the previous chapter (Fig. 3.1), the reconfigurable logic gates demonstrate interchangeable PUN and PDN based on a single reconfigurable input.

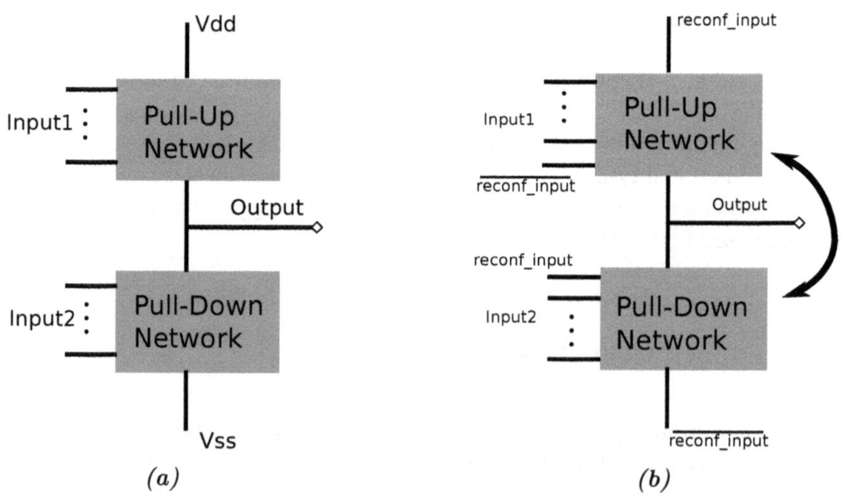

Fig. 4.1 (**a**) Fixed pull-up and pull-down network in case of complimentary MOS logic gates. (**b**) Interchangeable pull-up and pull-down network in case of RFET-based logic gates. The reconf_input decides the logic functionality

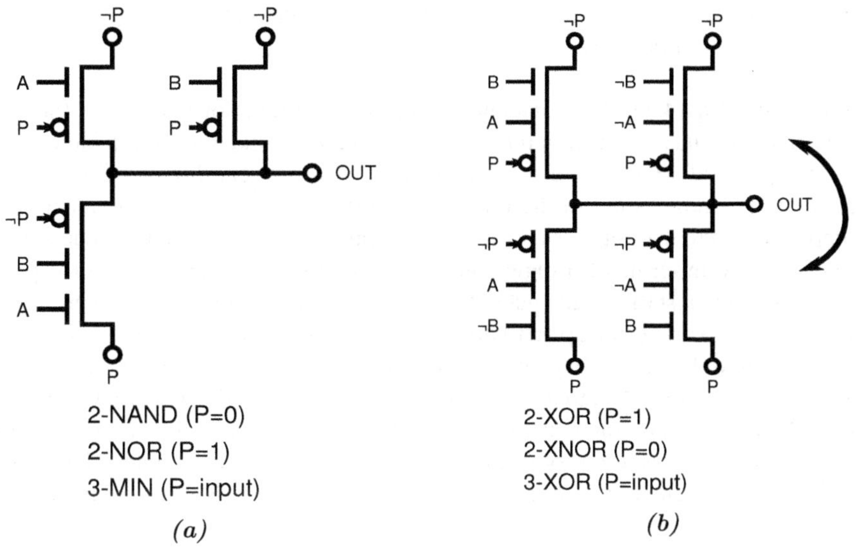

Fig. 4.2 Reconfigurable logic gate MINORITY and XOR3 demonstrated in [Hei+12, Rai+19]. It shows how functionality changes with the value of P. With the change in the value of P, pull-up and pull-down networks get interchanged

4.4.2 Reconfigurable Truth Table

The truth table of a logic gate represents the electrical characteristics of a given circuit schematic in terms of logic 1 and logic 0. For reconfigurable logic gates as shown in Fig. 4.2, it is shown in this chapter that the interchangeability between pull-up and pull-down networks can be well encapsulated in *self-dual* logic. The concept of *self-duality* is integral in understanding the reconfigurable truth table as it correctly abstracts interchangeable PUN and PDN of a logic gate.

Self-dual Functions In order to understand *self-dual* functions, it is necessary to understand what a dual of a function in Boolean space implies. The *dual* of any function $f(x_1, \ldots, x_n)$ is given by $\bar{f}(\bar{x}_1, \ldots, \bar{x}_n)$ and is denoted by f^d. The function f^d is obtained first by replacing each literal x_i with $\bar{x_i}$ and then by complementing the function.

A logic function $f(x_1, \ldots, x_n)$ is called *self-dual* [Sas12] if and only if

$$f(x_1, \ldots, x_n) = \bar{f}(\bar{x}_1, \ldots, \bar{x}_n), \tag{4.1}$$

or, equivalently, if

$$\bar{f}(x_1, \ldots, x_n) = f(\bar{x}_1, \ldots, \bar{x}_n) \tag{4.2}$$

for all $x_1, \ldots, x_n \in \mathbb{B}$.

Theorem 1 *All self-dual functions can be implemented as a reconfigurable logic gate using any single reconfiguration input.*

Proof From Eq. (4.1), it is apparent that in the truth table of a self-dual function, every maxterm has a dual in a minterm, such that the complete truth table consists of only pairs of dual maxterm and minterm. Thus, every literal of a maxterm has a corresponding dual in the minterm. Similarly, the output of the maxterm is complemented to the output of the corresponding dual minterm. When representing the truth table in terms of a logic gate, every variable of a truth table corresponds to a signal in a static logic gate. Since the maxterm (minterm) represents the pull-down network (pull-up network) of a static logic network, the self-duality relation can only hold when the pull-up network (pull-down network) is active for a particular set of signals that can activate the pull-down network (pull-up network) in their complemented form. Ambipolarity of RFETs allows dynamic reconfiguration between p-type and n-type behavior. This enables the pull-up network to dynamically reconfigure to pull-down network thus realizing self-duality. □

For a network of transistors (pull-up and pull-down network) to work, all the signals need to be flipped so that the pull-up and pull-down networks get interchanged in a symmetrical way. Such a scenario is easily possible with RFETs due to their inherent reconfigurability at the device level. It is to be noted that while self-dual logic gates are also realizable with CMOS technology, due to dynamic reconfigurability, RFETs implement self-dual logic with a less number of transistors. For example, a MAJORITY logic function can be realized with CMOS using ten transistors [WH15], but in case of RFETs, it can be realized with four transistors [AGM13].

Figure 4.3 shows an example of a truth table for a self-dual function, *Minority*. Here, when the truth table is divided over the value of *C* (or any other arbitrary literal

C	B	A	Output
0	0	0	1
0	0	1	1
0	1	0	1
0	1	1	0
1	0	0	1
1	0	1	0
1	1	0	0
1	1	1	0

= 1110

Split Truth Table are bitwise complementary. This implies reconfigurable pull-up and pull-down networks

= 0001

Fig. 4.3 Truth table for minority logic gate. The truth table is split over the value of C (or any other input) which is the reconfigurable input per se), the two parts of the output's truth table are dual to each other. Similarly, it can also be seen from the truth table that the minterms and maxterms are dual to each other and exist in pairs.

Trivial and Nontrivial Self-Dual Functions

Among two-input functions, self-duality exists in those functions which are equivalent to either the inputs or to their complements (for example, $f(a, b) = a$ or $f(a, b) = \overline{a}$). Such functions are implemented in circuits as wires (or used an inversion), and hence, their implementation in RFETs is identical to that of CMOS. Thus, two-input self-dual functions are referred to as *trivial functions* in this thesis. For self-dual functions with more than two inputs, their implementation with RFETs requires fewer transistors compared to their CMOS counterpart [Rai+19, Rai+20]. These functions will have a direct impact on the area of the circuit. Hence, self-dual functions with 3 or more inputs are referred to as *nontrivial functions*.

4.5 Distilling Standard Cells

In this section, an approach is presented to extract possible standard cells from a given circuit. Algorithm 4.1 detects self-dual cuts that can be implemented as reconfigurable logic gates. The algorithm traverses through a circuit and observes the number of specific cuts in a logic network that are self-dual. The algorithm is applied during technology mapping of a network of a circuit using mockturtle [Soe+18].

The input to Algorithm 4.1 is a k-input cut list for a particular benchmark, where k is the number of inputs in each cut. Each cut is then iterated to see whether it is self-dual or not. If the condition is true, we maintain a hash table, $\mathbb{H}$ for each such truth table to count its occurrence in the circuit under test. This is done over a benchmark suite. This helps in profiling a particular circuit in terms of the Boolean functionality required at individual cuts. Additionally, we maintain a *hash table*, $\mathbb{F}$ which contains the smaller functions possible with the reconfigurable logic gate. This is important to understand what kinds of reconfigurable functions are possible.

Since nontrivial self-dual functions are required, only three- or more-input logic gates are extracted since the CMOS-based implementation for such distilled reconfigurable logic gates is not efficient in terms of the number of transistors and performance [Rai+19]. Hence, having these logic gates available in the RFET-based generic library can potentially improve the technology mapping of circuits. The count of each entry of the hash table is also maintained to get the total number of functions used.

Algorithm 4.1 Finding standard cells which are reconfigurable

Data: k-input cut list for a benchmark G
Result: hashTable, $\mathbb{H}$ consisting of self-dual standard cells
Result: hashTable, $\mathbb{F}$ consisting of Shannon decomposition of functions in $\mathbb{H}$
$cutList = cut\_enumeration(G)$ **foreach** cut, $c \in cutList$ **do**
  if $(c.is\_selfdual())$ && $(c.num\_var() \geq 3)$ **then**
    $pair\{f_1, f_2\} = c.compute\_cofactors()$ $\mathbb{H}.insert(c)$ $\mathbb{F}.insert(f_1, f_2)$
  end
end
return H

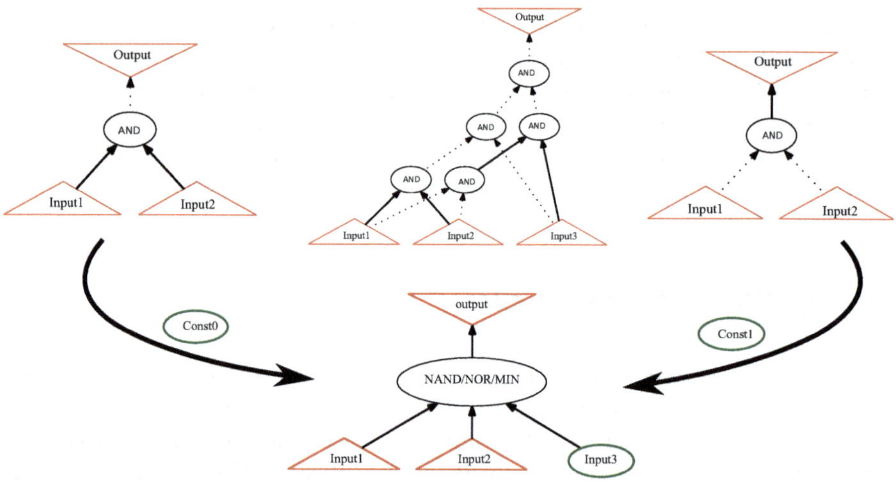

Fig. 4.4 HOF-based mapping. Three different AIGs using the same logic gate during mapping, albeit with the addition of constant logics. The third input can be const0, const1, or just another input for NAND, NOR, or MINORITY logic, respectively

4.6 HOF-Based Technology Mapping Flow for RFET-Based Circuits

Reconfigurability is achieved in RFETs at runtime by changing the polarity of the program gate terminals. In terms of mathematics, this functional reconfigurability can be encapsulated using a Higher-Order Function (HOF) as described in the following equation:

$$\begin{aligned} f(x, y, z, w) &= g(x, y, z), when\ w = 0 \\ &= h(x, y, z), when\ w = 1 \end{aligned}$$

Here f is an HOF of four variables. Functions g and h are two different functions. In the above expression, f can be represented in terms of functions g and h depending upon the values of w. Function f can be seen encapsulating functions *g* and *h*. Analogous to the above mathematical function, RFET has been shown to behave as a p-type or an n-type device, and that choice can be made by changing the potential of the PG. The PG is the *w* of the above function.

Extending this transistor behavior, we have logic gates that can show multiple functionalities from the same structure as mentioned in [Tro+16]. Such instances of logic gates that show multiple functionalities can be termed as HOF logic gates.

An important point to add here is that these *HOFs* can only give certain outputs depending upon a certain configuration of the inputs. Having an electrical garbage value from the misconfiguration of such logic gate is a valid possibility. Such misconfiguration can lead to security vulnerabilities, which is discussed in Chap. 7. HOF-based mapping can be explained as shown in Fig. 4.4. There are three different AIGs for different functions—NAND2, NOR2, or MINORITY. However, in the case of HOF-based mapping, all three AIGs are mapped onto a single HOF logic gate as shown in the figure. The concept of HOF-based mapping is novel in the context of RFETs as reconfigurable logic gates implementing multiple yet mutually exclusive functionalities are possible with a less number of RFET transistors. Such expressive power of intrinsic reconfigurability can be very well represented using *HOF* gates. Two kinds of mappings are possible with RFETs:

1. *Static mapping*: Here, static logic gates based on RFETs are used. The logic gates used in this mapping contain only single-function logic gates as implemented by RFETs. These logic gates are similar to CMOS logic gates with a distinct V_{dd} and V_{ss}.
2. *Reconfigurability-aware mapping*: Here, RFET-based reconfigurable logic gates are used. The mapping results in a circuit where some or all logic gates are reconfigurable. The mapping provides the area that considers the overhead involved using reconfigurable logic gates. Inverter contributions due to driving of signals to the V_{dd} and V_{ss} are considered.

4.6.1 Area Adjustments Through Inverter Sharings

An important difference to note among the logic gates proposed in [Tro+16] is the XOR family of logic gates. In both renditions of the XOR gate, i.e., *2-bit XOR* or *3-bit XOR*, the complemented and actual forms of each input are required in the logic gate. So to calculate the actual number of transistors involved in these logic gates, an inverter must be included for getting the inverted phase of each input within the logic gate boundary. Like in two-input XOR_XNOR, both inputs—*A* and *B*—are required in both phases. Hence, the overall area for the XOR-based logic gates should increase by a factor of $\#\, of\ input\_variables$ x 2 RFETs. This is an overhead that must be taken into account.

However, if the XOR logic is used at multiple places in a logic circuit and if some of the fan-ins are available in the inverted form within the logic circuit, then that can lead to the redundant use of inverters. Hence, the feasibility of harnessing such inverted forms of fan-ins available in the circuit needs to be explored. An obvious problem arises as complemented forms of fan-ins available from other parts of the logic circuit add to the fan-out delay due to the longer length of metal wires. This problem can be solved by utilizing the unique electrical properties of multi-input RFETs. The XOR family of logic gates uses interior gate terminals of multi-input RFETs. Authors in [Rai+17] state that the gate terminals placed above the Schottky barrier in RFETs are faster gate terminals. In Fig. 2.1a, I–V characteristics of interior gate terminals exhibit a steeper slope for transitions. In XOR both these inputs are fed to these faster gate terminals. Therefore, the faster interior gate terminals compensate for the delay caused by the long length of metal wires. Hence, in cases of logic circuits that have these inputs already available in inverted forms in other parts of the circuit, it is feasible to use such inverted forms for these logic gates thereby reducing the overall number of transistors required in the circuit.

Figure 4.5 illustrates area optimization in logic network. The cloud-shaped parts represent the combinational parts of a logic network. There are three XOR nodes whose fan-ins are shown along with a bubble, signifying inverters required for complemented input. Consider the XOR node 1 that has one fan-in coming from the primary input (blue arrow) and the other coming from the left combinational part (red arrow). However, it can be seen that inverted forms of both the fan-ins

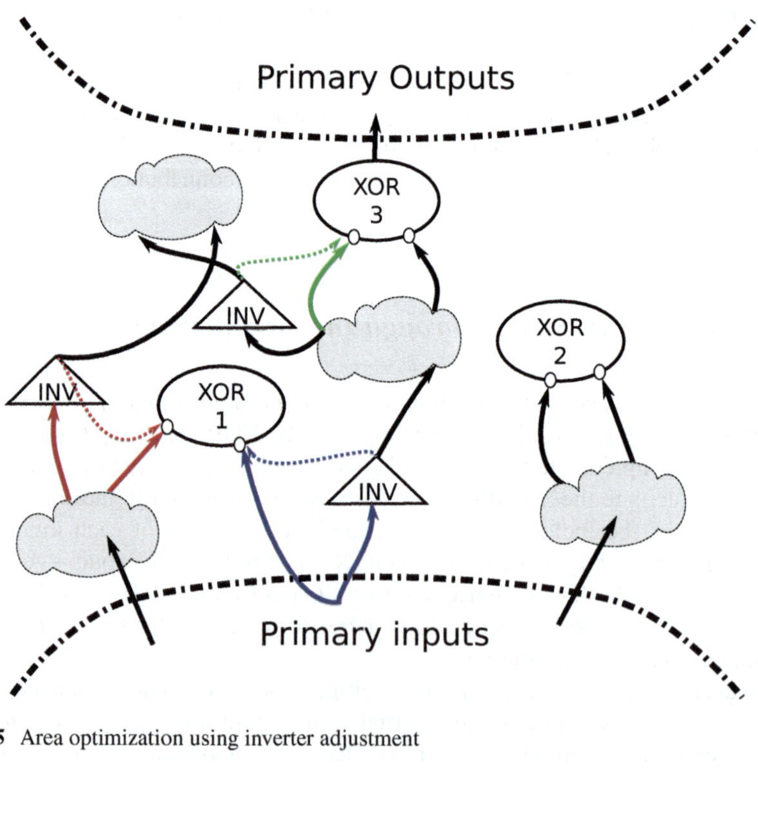

Fig. 4.5 Area optimization using inverter adjustment

are available in the circuit, and hence, a suitable scenario for area optimization is possible here as shown through dotted lines. In this case, the algorithm gets away with both the internal inverters in the XOR gate. Similarly, for XOR node 3, one of the fan-ins are available in inverted form (green arrow) and that too contributes to area saving. XOR node 2 has none of the fan-ins available from the circuit, and hence, inverter contribution for both the fan-ins has to be added in the final area calculation. Logic networks using the XOR-based logic family can tap this kind of optimization to reduce the overall area. Further, limiting the inverter sharing for a certain number of fan-ins is required to prevent higher fan-out delays, if a single inverter output is shared by multiple XOR gates. Such logic sharing during the technology mapping also eases out later stages of physical design.

Algorithm 4.2 Area savings using already available inverted fan-ins

```
Data: Mapped Network G and Area
Result: Mapped Network G and updated Area Area
foreach node, n ∈ G do
    if n.is_XORfamily() then
        foreach fanin, f ∈ n do
            if f.is_complemented()  &&  G.is_node(f) then
                f.mark_for_inv_adjustment()
            end
        end
    end
end
return Area
```

To calculate the number of such available inverted signals, the pseudo-code shown in Algorithm 4.2 is used. The initial area during mapping is calculated in *Area*. If a particular node matches with the XOR family, available inverted forms of fan-ins are searched in the overall circuit. An area adjustment is required at the same time for every available fan-in. The mapper finally gives that mapping, which has the lowest area of the overall circuit. Instances of XOR logic gates that do not find such available fan-ins contribute to their default area.

4.6.2 Technology Mapping Flow

Figure 4.6 shows the complete flow of our technology mapping for logic design based on reconfigurable FETs. The output after the logic optimization is fed to our technology mapper. The Higher-Order Function encapsulates reconfigurable logic gates based on RFETs. At first, the mapping is carried out using HOF followed by typical area optimization flows in a technology mapper [Cha07]. The output of HOF-based technology mapping is a netlist containing reconfigurable-aware logic gates.

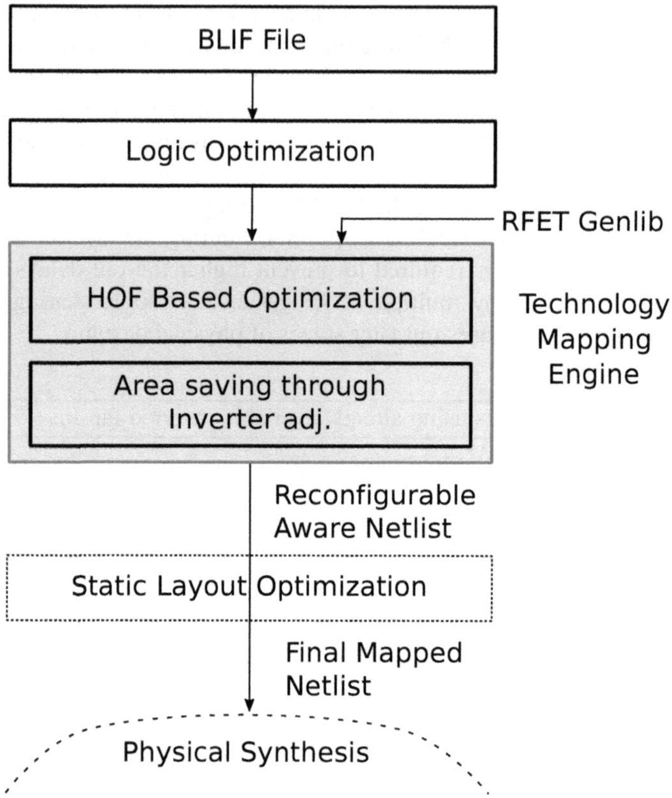

Fig. 4.6 Entire technology flow suited for RFETs

4.6.3 Realizing Parameters for the Generic Library

While populating the generic library, the number of transistors is considered for area calculations as that is consistent with logic synthesis for CMOS technology as well. For experiments, all types of RFETs contribute as a unit transistor to the area. For the logic gates distilled using the method proposed in Sect. 4.5, the area is computed using the following method. The Boolean expression for all the logic gates is first represented in an Sum-of-Products form. For each minterm consisting of literals that are AND-ed together, a single RFET is considered. This is based on the fact that RFETs can support multiple independent gates on a single channel due to their wired-AND property [Sim+18]. Hence, a single RFET can have up to 4 control gate inputs with 1 program gate input. For the present work, up to 5-input standard cells are only considered since most of the technology mapping algorithms [Cha07, Cal+22] consider a cut size of up to 5 inputs in their default flow. Several works like [Zha+14, Tro+16, Tro+15] have showcased 4- to 5-input logic gates but none of the other works showcased a 6-input logic gate based on RFETs. Arguments can be

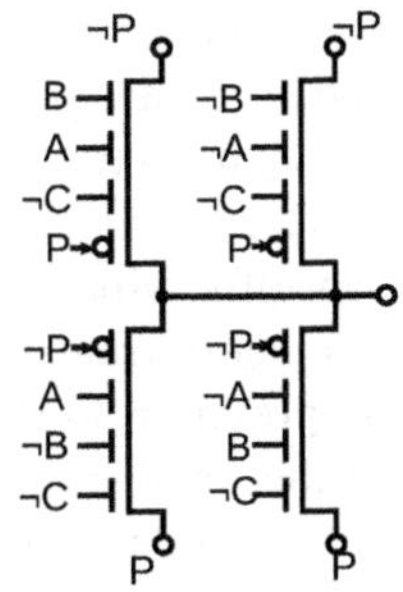

Fig. 4.7 Realizing parameters for a given function

made that RFETs support multiple independent gates on a single channel, but for the present scope, ease of understanding, and to be consistent with the literature, 6-input logic gates are not considered. Hence, parameters of up to 5-input reconfigurable logic gates can be easily calculated. Further, for logic gates, which are binate (for example, XOR3) in their literals, an additional inverter is considered for every binate literal while calculating the total number of transistors in a standard cell.

Let us take an example of a 4-input function represented by truth table 0000_1001_0110_1111. The Boolean expression can be represented as $f = !d * (!c * (!a*!b + a * b)) + d * (!c + (a*!b+!a * b))$ Here, d is the reconfiguration input. From the logic gate schematic, it is a simple extension of a 3-XOR logic gate as shown in Fig. 4.7. The first term consists of two minterms, while the second term consists of three minterms. So, the number of transistors is 5 (2 in pull-up and 3 in pull-down) to implement these minterms. Additionally, since all the variables are binate, this further requires four additional inverters. So, the total number of RFETs for implementing the above function is $5 + 4 * 2 = 13$. Here, the contribution of all individual RFETs (irrespective of whether they are TIGFET or MIGFET) in terms of area is considered as 1 transistor. For delay calculation, the number of logic levels between the primary input to primary output is used. This corresponds to the depth of a logic network.

4.6.4 Defining RFET-Based Genlib for HOF-Based Mapping

To support multiple functionalities for RFET logic gates, the generic library (*genlib*) has to be modified to support the abovementioned changes. Since RFET has multiple functionalities, a logic gate entry in genlib has to be updated with the number of outputs it can support and a description of those outputs. The sample generic library is shown in Table A.1 (Appendix A) In order not to interfere with the normal flow of ABC, the `sinw` identifier is used in the genlib. The area is considered in terms of the number of transistors. Hence, an inverter has 2 RFETs, and just like the CMOS flow, the area of the inverter is normalized to 1. The area of all other logic gates is normalized with respect to the area of the inverter. To enable the area optimization due to inverter sharing in the ABC tool, an additional `inv_adj` identifier has been added in the genlib to the gate definition.

Let us to try to understand with an example from Table A.1 in Bijlage A. For the logic gate *nand_nor_min* the number 3 is the number of functions this logic gate can offer which are defined by $O1$, $O2$, and $O3$. Then 2.5 is the area in terms of number of transistors normalized to the area of an inverter. In [Rai+17], the authors proposed an E-MUX to replace a two-stage MUX with a single logic gate which further enabled reduced area and delay numbers. The *E-MUX* logic gate has also been added in our genlib. This approach provides a simple mechanism to include and analyze the benefits of such futuristic combinational logic gates based on RFETs.

4.7 Experiments

The whole experiment section is divided into two major parts. In the first part (4.7.1), Algorithm 4.1 is used to list out the major self-dual (reconfigurable) logic gates present in a benchmark suite. For this analysis, the EPFL benchmark suite [AGD15] is used. However, the approach can be extended to any kind of benchmark suite available. The *cut-enumeration* method implemented within mockturtle [Soe+18] is used to enumerate the varying k-input cuts for individual benchmarks where k is $3, 4, 5, 6$, respectively.

In the second part, two mappings are evaluated. First, HOF-based mapping is evaluated that indicates how mapping with RFET-based logic gates gives a clear advantage over CMOS-styled RFET-based logic gates in terms of area (number of transistors), delay, edge, and runtime. This empirically demonstrates the findings of the previous chapter at a broader benchmark level. Both static and reconfigurable flows are shown and the results are analyzed.

In the second mapping, the default standard-cell-based mapping is carried out to analyze the effect of newly *distilled* standard cells on the overall area, delay, and edge connections of individual benchmarks. Here, the self-dual logic gates found in the first experiment are used as standard cells to constitute a generic library. While the first HOF-based mapping is done to demonstrate the new mapping flow, the second mapping is carried out to demonstrate the impact of the newly distilled standard cells.

4.7.1 *Experiment 1: Distilling Standard Cells from a Benchmark Suite*

Table B.1 in Appendix B shows the overall occurrences of such reconfigurable logic gates across all the benchmarks for 3-, 4-, and 5-input cut mapping. This corresponds to *distillation* of an equivalent number of input logic gates. A 6-input cuts are not considered. This is in coherence with the multi-input gate terminal feature as explained in Sect. 2.1.3. It is to be noted that the feature of multi-input gate terminals in RFETs prevents cascading of transistors in a single path. Cascading of transistors, in general, is not favorable for standard cells as it leads to slower logic gates. Hence, more than 3-input logic gates are generally avoided in contemporary CMOS technologies.

Algorithm 4.1 returns in a total of 132 self-dual function for EPFL benchmarks as shown in Table B.1. The complemented form of the already distilled truth table is ignored as the ABC mapper [BM10] already takes care of complemented forms during technology mapping. For example, logic gates denoted by truth tables 00010111 (*Minority*) and 11101000 (*Majority*) are equivalent as one can be implemented by the other just by adding an inverter. However, looking at the number of occurrences, it can be observed that 65 functions occurred more than ten times for EPFL benchmarks. Appendix B also contains the self-dual functions found and the two smaller functions that are dual to each other. Apart from the popular functions such as XOR3 and NAND3-NOR3, several other interesting functions of 4 and 5 variables are also reported. This allows the user to specifically select a list of RFET-based standard cells.

In the next set of experiments, both mapping approaches are evaluated subsequently.

4.7.2 Experiment 2A: HOF-Based Mapping

Here, the HOF-based mapping along with inverter adjustment flow is evaluated. The optimization for mapping with HOF-based logic gates is done in the ABC tool. All the standard optimizations in ABC are intact and there is no interference with the default flow. The ABC command of `strash` and `balance` are used before calling the `map` command on a benchmark circuit. For runtime calculations, `time` command after `map` command is invoked. It is to be noted that delay-oriented mapping has been used for all experiments. The area, delay, and runtime values for each benchmark are compared for the EPFL benchmark suite. Here, circuits based only on RFETs are considered and comparison is carried among the three flows which use only RFET-based logic gates. Three types of genlib are used:

1. For the baseline, genlib consisting of CMOS-styled logic gates based on RFETs are used. Here, the number of transistors in each gate is similar to the one available in CMOS-based generic library. One-to-one substitution of CMOS transistors (both PMOS and NMOS) is done with RFETs.
2. For static genlib, logic gates with just one functionality is used (Table 3.1). From the layout point of view, logic gates maintain the Vdd and Vss instead of using inverters for generating P and $\overline{P}$. However, with this change, the reconfigurability of the logic gate is lost.

3. For mapping, reconfigurable logic gates as described in the previous chapter (Table 3.2) are used. Here, individual logic gates can have more than one functionality. An additional inverter area has to be considered for the logic gate to steer P and $\overline{P}$ in this mapping.

Runtime Comparison

The comparison in runtime for the three flows is shown in Fig. 4.8. The runtimes are measured in seconds. From the graph in Fig. 4.8, it is apparent that the baseline flow is the fastest of three flows. This can be attributed to the time taken by the inverter adjustments in static and reconfigurable flows. The runtime overhead for the two flows can be further correlated with Table 4.1. For example, in the case of *log2* benchmark, one can notice that the area savings due to inverter adjustment in the static flow are more, and hence, the runtime is also more. For large benchmarks with a higher possibility of inverter adjustments (Table 4.1), the static flow on average has the largest runtime, followed by the reconfigurable flow and then the baseline flow.

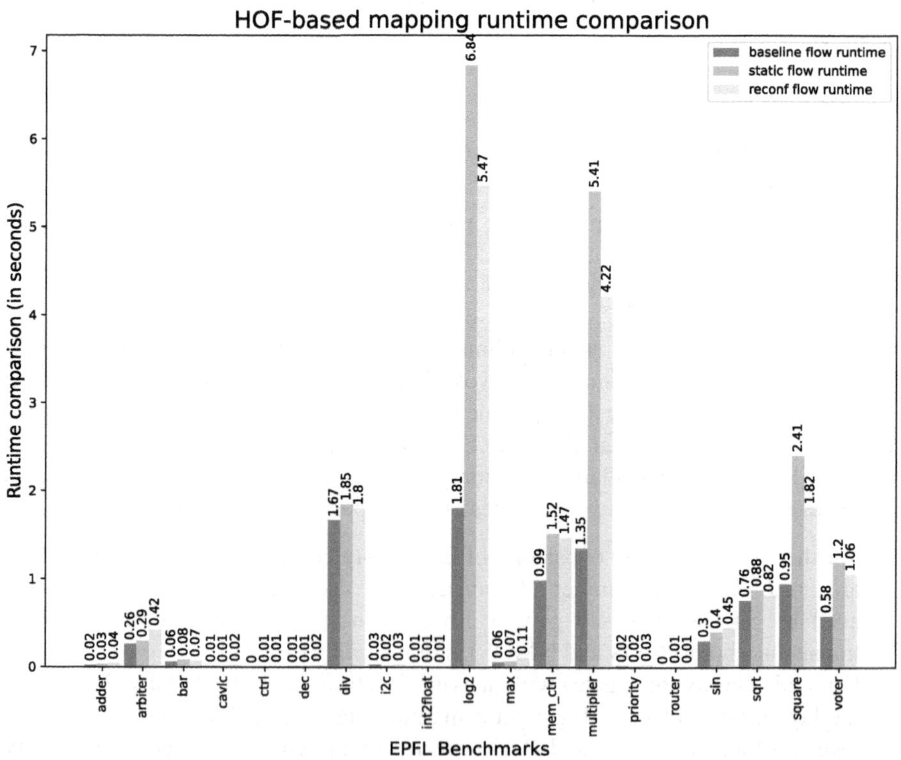

Fig. 4.8 Runtime comparison. The reconfigurable and static flow has a higher runtime as compared to the baseline flow due to inverter adjustments

This is due to the fact, as in the case of static flow, that the genlib contains more gates as compared to the reconfigurable flow. Hence, the mapper has to iterate over more logic gates as compared to the reconfigurable flow. The final runtime averages for the baseline, static, and reconfigurable flows are 0.47, 1.10, and 0.94 s, respectively.

Area Comparison

The comparison in area values is shown in Fig. 4.9. Here, the area values for both the static and the reconfigurable flows are normalized with respect to the baseline area values. It is to be noted that the area savings due to inverter adjustments are not deducted from the area values and is shown separately in Table 4.1. The first thing that can be noticed is that both the static and reconfigurable flows using RFET-styled gates achieve better area values as compared to the baseline flow. However, few exceptions can be seen for reconfigurable flow particularly for small benchmarks such as *cavlc*, *dec*, and *int2float*. The area is even more than that of

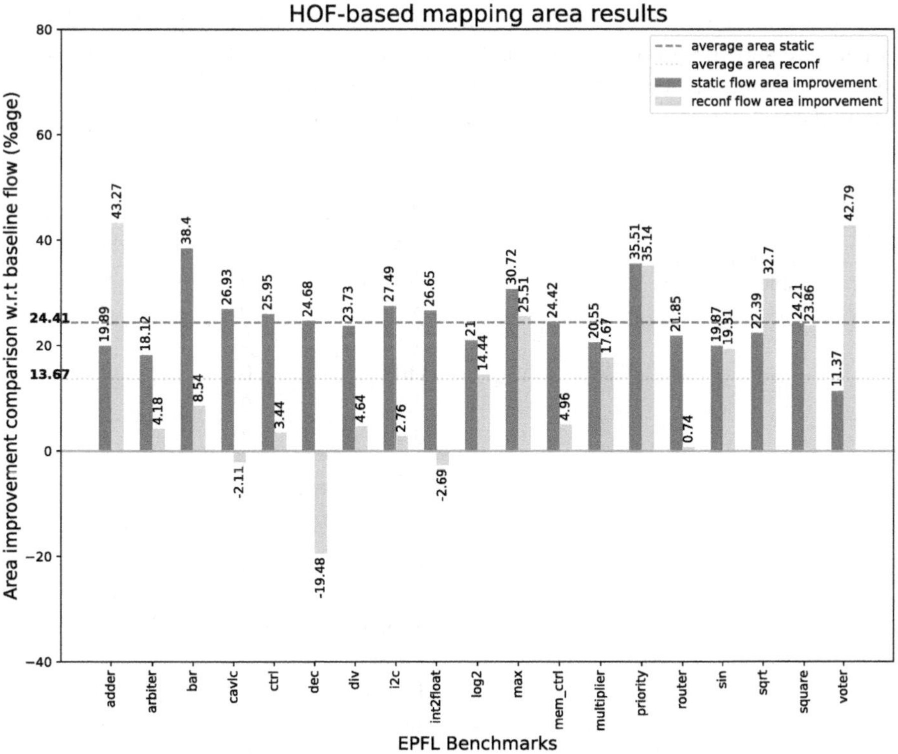

Fig. 4.9 Area comparison for various flows. Both static and reconfigurable flows achieve lower area as compared to the baseline flow. The static flow is the best in terms of area as average area improvement over the baseline is 24.43%

the baseline flow. This is due to the fact that in reconfigurable flow, the smaller logic gates are not available for mapping as an additional inverter is required for reconfigurability. Hence, the area values for these benchmarks are worse across the three flows. Further, the static flow achieves the least area number as compared to the other two flows for all the benchmarks. The average improvements across static and reconfigurable flows over the baseline flow are 24.43% and 13.67%, respectively.

The area savings due to inverter adjustments for both static and reconfigurable flows are shown in Table 4.1. It can be seen that not all circuits show inverter adjustments as only circuits using the XOR family of gates show better area savings with inverter adjustments. For example, in the case of *voter* benchmark, the number of instances of XOR gate usages in static and reconfigurable flows are 2044 and 1101, respectively. Hence, there is such a large discrepancy in the area savings due to inverter adjustments.

Delay Comparison

In this experiment, for every RFET-based logic gates in the generic library, a unit delay is considered. The delay comparison is shown in Fig. 4.10. Here, also the values are shown with respect to the baseline values. The delay value refers to the number of logic levels between the primary input and the primary output as all logic gates have unity delay in the generic library of all three flows. In terms of

Table 4.1 Comparison in the area savings due to inverter adjustments for static and reconfigurable logic gates

Benchmarks	inv adj static	inv_adj reconf
Adder	191	128
Arbiter	0	0
bar	0	0
cavlc	2	0
ctrl	0	0
dec	0	0
div	7	4
i2c	0	0
int2float	0	0
log2	2635	2355
max	0	0
mem_ctrl	166	169
Multiplier	2935	2698
Priority	0	0
Router	19	17
sin	340	235
sqrt	0	0
Square	1309	1245
Voter	1346	113

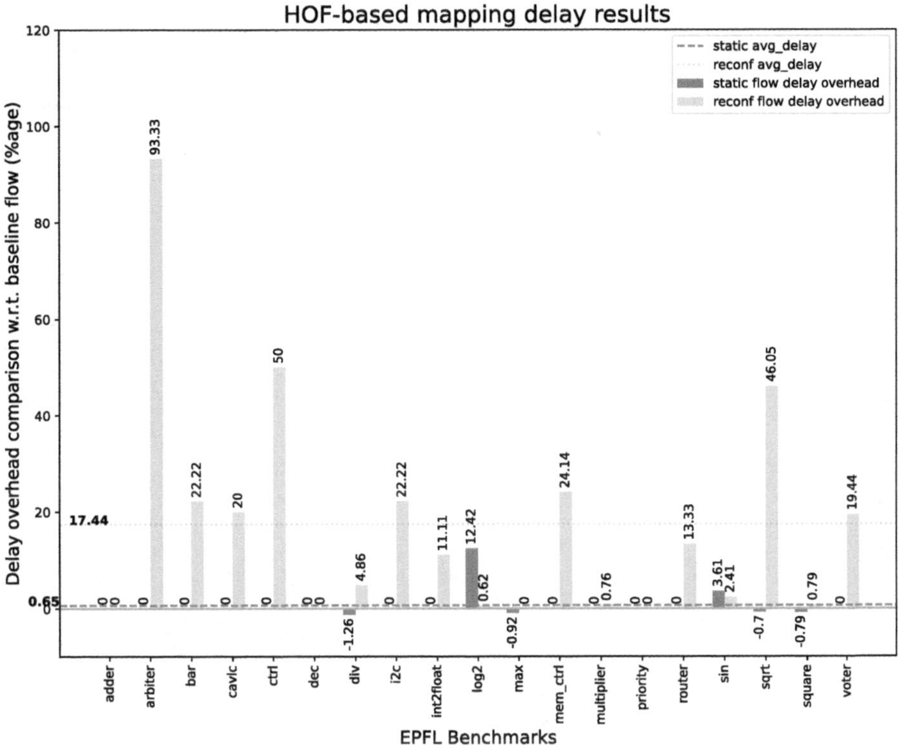

Fig. 4.10 Delay comparison for various flows. In terms of delay overhead, the reconfigurable flow has the largest overhead. This is primarily because reconfigurable flow uses larger logic gates with more inputs as compared to the other two flows

delay, the reconfigurable flow has the largest delay with an average overhead of around 17.44% over the baseline flow because the reconfigurable flow uses a lot of reconfigurable logic gates which necessitates the use of constant logic such as *const0* or *const1*. This is particularly worse for benchmarks such as *arbiter* as other flows just use only NANDs or NORs, while the reconfigurable flow uses the bigger MINORITY logic gate. On the other hand, the static flow has a small overhead of just 0.65% as compared to the baseline flow.

Edge Comparison

The number of edges refers to the total number of connections within the circuit and correlates with the wirelength obtained post-physical synthesis. The comparison in the number of edges for the three flows is shown in Fig. 4.11. Both the static and reconfigurable flows have on average more edges (0.60% and 3.65%, respectively) as compared to the baseline flow. As discussed before in delay comparison, this

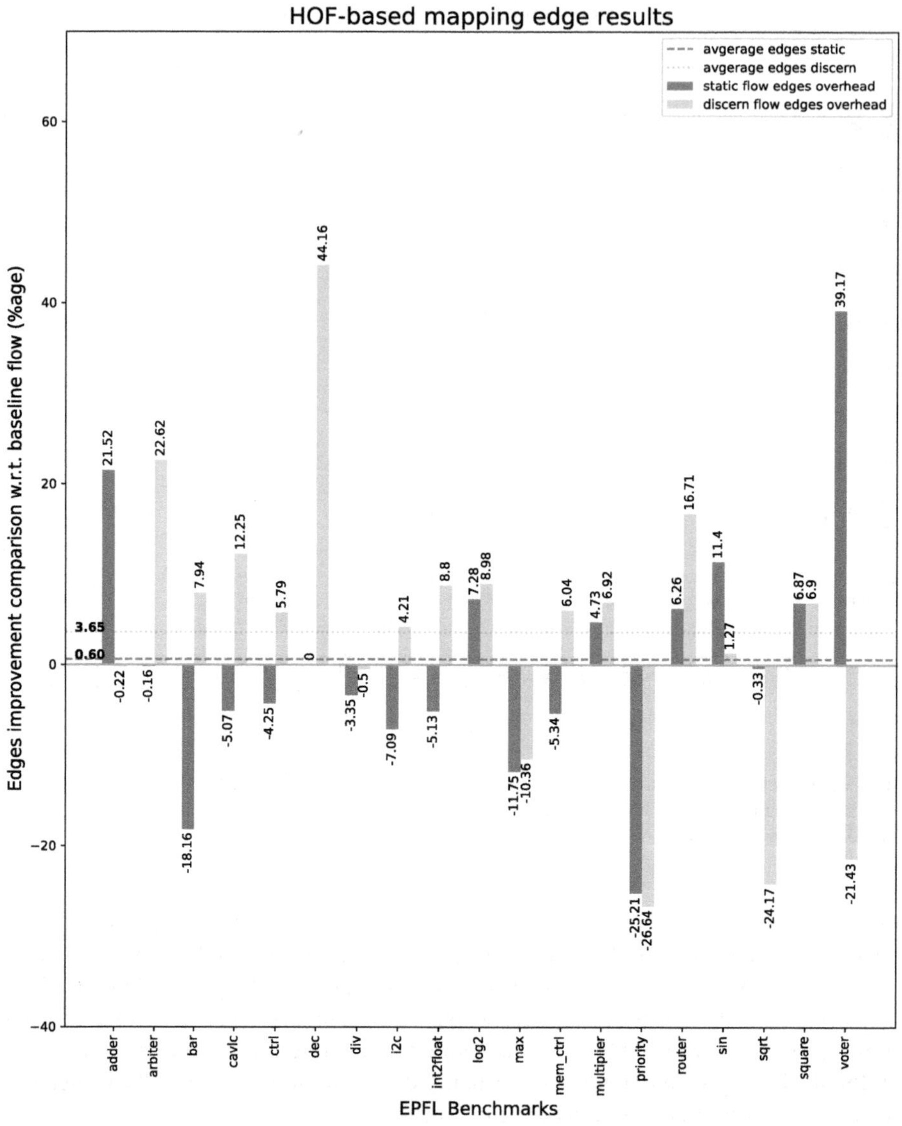

Fig. 4.11 Edge comparison for various flows. As compared to the baseline, there is a marginal increase in the number of edges for the other two flows

increase in the number of edges is obvious as larger logic gates are being used for smaller gates that require more number of constant logic 0 or constant logic 1. For example, if a MINORITY is used in place of 2-NAND or 2-NOR, then constant logics are required during mapping.

The HOF-based mapping aims to provide the complete spectrum in physical parameters of area, delay, edges, and runtime for circuits in terms of RFET-based mapping. It presents both the improvements and overheads of RFET-based mapping where logic gates can be used either with static functionality or reconfigurable functionality.

4.7.3 *Experiment 2B: Using the Distilled Standard Cells During Mapping*

The previous experiment as described in Sect. 4.7.1 provides a list of reconfigurable standard cells based on the functional profiling of a benchmark suite. In this mapping experiment, these newly distilled standard cells are used to evaluate the impact they have on the circuit's parameters like area, delay, and the number of edges.

Just like in the HOF-based mapping, three different generic libraries are also used for this experiment. The first and second one are the same as used in the previous experiment 2A, i.e., the baseline flow (library-A) and the static flow (library-B). The third generic library is the superset of the library-B with few of the newly distilled standard cells as found in the first set of experiment (listed in Table B.1). Mostly reconfigurable logic gates include *impl-nonimpl*, *aoioai21*, *aoioai22*, *nand3-nor3*, and two 5-input logic gates (from Table B.1, logic gates with high occurrence are added). However, this selection is arbitrary and is also benchmark suite dependent.

Mapping Using Standard Cells Through Circuit-Level Excavation

The absolute improvement over the baseline flow for area, delay, and edge comparison for EPFL benchmarks are shown in Figs. 4.12, 4.13, and 4.14, respectively. The runtime for all three flows is similar to the average runtime across all the benchmarks falling in the range of [0.46, 0.48] s. In terms of area improvement over the baseline, mapping with library-B achieves 25.47%, while library-C achieves 29.59% average area improvement, respectively.

In terms of delay, all flows achieve similar results with library-C achieving the best delay results across all flows. All the benchmarks show a nonnegative delay improvement with library-C. Similarly, in terms of edges, library-C again delivers the best results with an average improvement of 6.37%. With this result, it can be observed that with appropriate RFET-based standard cells, the total number of edges can be reduced. Except for the *router* benchmark, all other benchmarks have achieved improvement in the number of edges.

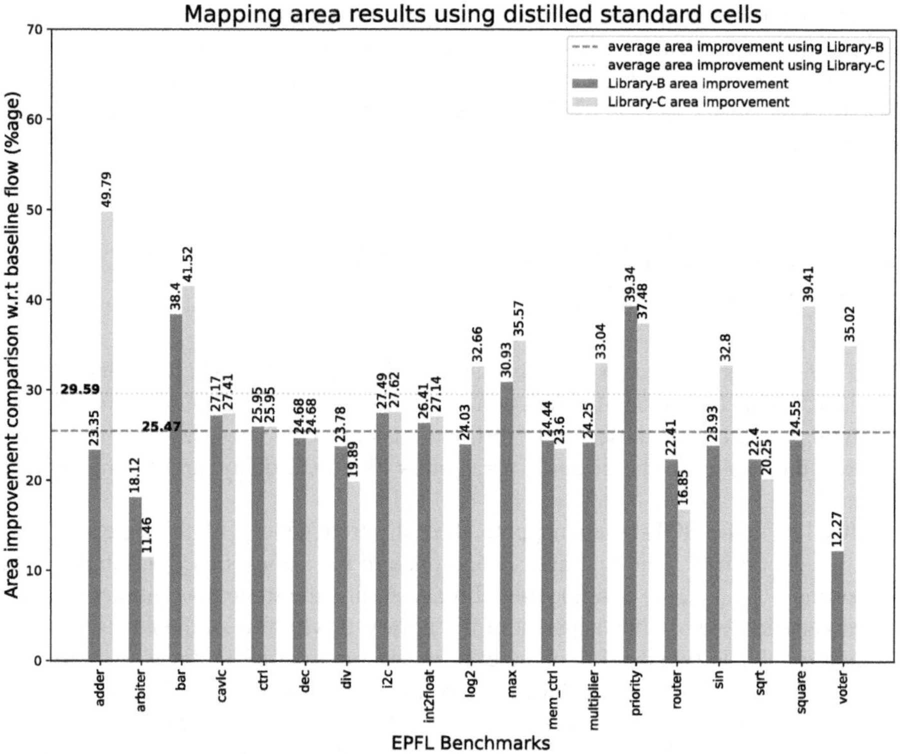

Fig. 4.12 Area comparison for various flows. It can be seen that as compared to the baseline, library-C achieves the best area improvement

4.8 Concluding Remarks

This chapter presents an EDA-centric approach towards enabling RFET-based circuits. First, a logic abstraction for the RFET-based logic gate is presented. It has been shown that self-dual Boolean logic gates can efficiently encapsulate duality offered by individual RFETs and are the natural logic abstraction for RFET-based logic gates. Then, an algorithmic approach is proposed that can be used to distill RFET-centric standard cells by analyzing self-dual cuts in a benchmark suite. This approach is used to profile various standard cells and their occurrence across a benchmark suite.

Further, mapping approaches that can exploit the reconfigurability offered by RFETs during technology mapping have been demonstrated. The concept of higher-order functions has been elucidated to represent RFET-based logic gates and to use them during technology mapping. Further, a mechanism to contribute to area saving was presented for the XOR-based logic family using already available inverted forms of fan-ins. Through experiments, a comparison between static and

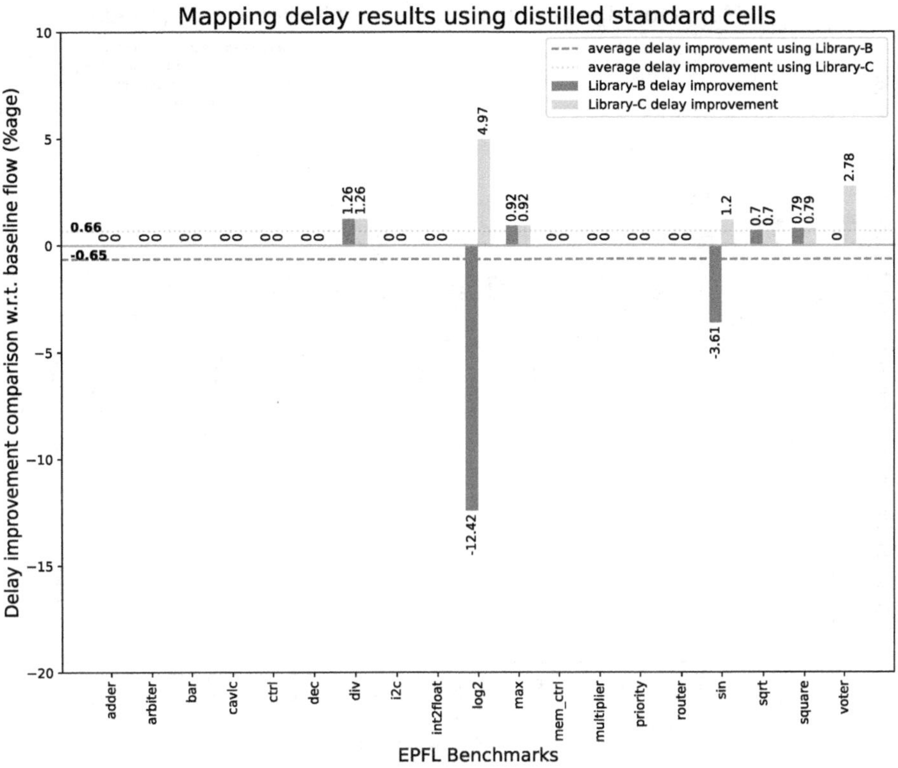

Fig. 4.13 Delay comparison for various flows. With library-C, the delay improvement of 0.66 is achieved as compared to the baseline flow

reconfigurable-ready technology mapping was done in terms of area, delay, edges, and runtime keeping the logic optimization and minimization steps intact. While both the static and reconfigurable flows show better area numbers, the reconfigurable flow shows overhead in terms of delay and number of edges. This observation is obvious considering the reconfigurable flow uses larger logic gates and also because of the use of more number of constant logic.

Last, through experiments over EPFL benchmark suites, using newly distilled logic gates (or self-dual logic gates) as standard cells, improvements in the area, delay, and edge are shown for EPFL benchmarks suites. It was shown that using such logic gates, best area, delay, and the number of edges can be achieved. This experiment demonstrates the impact of using self-duality as a natural Boolean abstraction for RFET-based circuits. An important takeaway from these experiments is that CMOS-styled logic gates may not be suitable for RFETs. Logic gate designs exploiting the duality of individual transistors can lead to efficient standard cell designs based on RFETs.

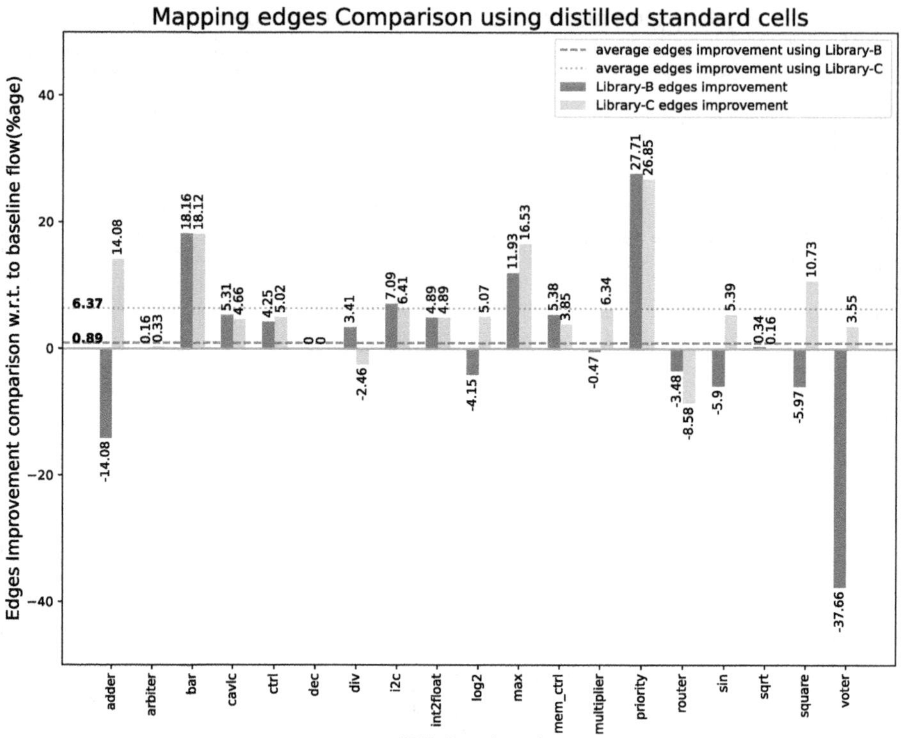

Fig. 4.14 Edge comparison for various flows. Both library-B and library-C show an average improvement of 0.89% and 6.37%, respectively, over the baseline

References

[AGD13] Luca Amarú, Pierre-Emmanuel Gaillardon, and Giovanni De Micheli. "Biconditional BDD: A Novel Canonical BDD for Logic Synthesis Targeting XOR-rich Circuits". In: *Proceedings of the Conference on Design, Automation and Test in Europe*. DATE '13. Grenoble, France: EDA Consortium, 2013, pp. 1014–1017.

[AGD14a] Luca Amarú, Pierre-Emmanuel Gaillardon, and Giovanni De Micheli. "An Efficient Manipulation Package for Biconditional Binary Decision Diagrams". In: *Proceedings of the Conference on Design, Automation & Test in Europe*. DATE '14. Dresden, Germany: European Design and Automation Association, 2014, 296:1–296:6.

[AGD14b] Luca Amarú, Pierre-Emmanuel Gaillardon, and Giovanni De Micheli. "Majority-Inverter Graph: A Novel Data-Structure and Algorithms for Efficient Logic Optimization". In: *Proceedings of the 51st Annual Design Automation Conference*. DAC '14. San Francisco, CA, USA: ACM, 2014, 194:1–194:6. https://doi.org/10.1145/2593069.2593158.

[AGD15] Luca Amarú, Pierre-Emmanuel Gaillardon, and Giovanni De Micheli. "The EPFL combinational benchmark suite". In: *Proceedings of the 24th International Workshop on Logic & Synthesis (IWLS)*. CONF. 2015.

[AGM13] L. Amarù, P. E. Gaillardon, and G. De Micheli. “Efficient arithmetic logic gates using double-gate silicon nanowire FETs”. In: *2013 IEEE 11th International New Circuits and Systems Conference (NEWCAS)*. June 2013, pp. 1–4. https://doi.org/10.1109/NEWCAS.2013.6573572.

[AGM16] L. Amarú, P. E. Gaillardon, and G. De Micheli. “Majority-Inverter Graph: A New Paradigm for Logic Optimization”. In: *IEEE Transactions on Computer-Aided Design of Integrated Circuits and Systems* 35.5 (May 2016), pp. 806–819. https://doi.org/10.1109/TCAD.2015.2488484.

[Ama+15] L. Amarú, P. E. Gaillardon, S. Mitra, and G. De Micheli. “New Logic Synthesis as Nanotechnology Enabler”. In: *Proceedings of the IEEE* 103.11 (Nov. 2015), pp. 2168–2195. https://doi.org/10.1109/JPROC.2015.2460377.

[BM10] Robert Brayton and Alan Mishchenko. “ABC: An academic industrialstrength verification tool”. In: *Computer Aided Verification*. Springer. 2010.

[Cal+22] Alessandro Tempia Calvino, Heinz Riener, Shubham Rai, Akash Kumar, and Giovanni De Micheli. “A Versatile Mapping Approach for Technology Mapping and Graph Optimization”. In: *2022 27th Asia and South Pacific Design Automation Conference (ASP-DAC)*. 2022, pp. 410–416. https://doi.org/10.1109/ASP-DAC52403.2022.9712552.

[Cha07] S. Chatterjee. *On algorithms for technology mapping (Doctoral Thesis)*. University of California, Berkeley, 2007.

[Hei+12] André Heinzig, Stefan Slesazeck, Franz Kreupl, Thomas Mikolajick, and Walter M. Weber. “Reconfigurable silicon nanowire transistors”. In: *Nano Letters* 12.1 (2012). PMID: 22111808, pp. 119–124. https://doi.org/10.1021/nl203094h. eprint: http://dx.doi.org/10.1021/nl203094h.

[Mar+12] M. De Marchi, D. Sacchetto, S. Frache, J. Zhang, P. E. Gaillardon, Y. Leblebici, and G. De Micheli. “Polarity control in double-gate, gate-all-around vertically stacked silicon nanowire FETs”. In: *2012 International Electron Devices Meeting*. Dec. 2012, pp. 8.4.1–8.4.4. https://doi.org/10.1109/IEDM.2012.6479004.

[Rai+17] M. Raitza, A. Kumar, M. Völp, D.Walter, J. Trommer, T. Mikolajick, and W. M. Weber. “Exploiting transistor-level reconfiguration to optimize combinational circuits”. In: *Design, Automation Test in Europe Conference Exhibition (DATE), 2017*. Mar. 2017, pp. 338–343. https://doi.org/10.23919/DATE.2017.7927013.

[Rai+19] S. Rai, J. Trommer, M. Raitza, T. Mikolajick, W. M. Weber, and A. Kumar. “Designing Efficient Circuits Based on Runtime-Reconfigurable Field-Effect Transistors”. In: *IEEE Transactions on Very Large Scale Integration (VLSI) Systems* 27.3 (Mar. 2019), pp. 560–572. https://doi.org/10.1109/TVLSI.2018.2884646.

[Rai+20] S. Rai, M. Raitza, S.S. Sahoo, and A. Kumar. “DiSCERN: Distilling Standard-Cells for Emerging Reconfigurable Nanotechnologies”. In: *DATE*. 2020.

[Sas12] T. Sasao. *Switching theory for logic synthesis*. Springer Science & Business Media, 2012.

[Sim+18] M. Simon, J. Trommer, B. Liang, D. Fischer, T. Baldauf, M. B. Khan, A. Heinzig, M. Knaut, Y. M. Georgiev, A. Erbe, J. W. Bartha, T. Mikolaiick, and W. M. Weber. “A wired-AND transistor: Polarity controllable FET with multiple inputs”. In: *2018 76th Device Research Conference (DRC)*. June 2018, pp. 1–2. https://doi.org/10.1109/DRC.2018.8442159.

[Soe+18] Mathias Soeken, Heinz Riener, Winston Haaswijk, Eleonora Testa, Bruno Schmitt, Giulia Meuli, Fereshte Mozafari, and Giovanni De Micheli. “The EPFL logic synthesis libraries”. In: *arXiv preprint arXiv:1805.05121* (2018).

[Tro+15] Jens Trommer, André Heinzig, Tim Baldauf, Stefan Slesazeck, Thomas Mikolajick, and Walter M. Weber. “Functionality-Enhanced Logic Gate Design Enabled by Symmetrical Reconfigurable Silicon Nanowire Transistors”. In: *IEEE Transactions on Nanotechnology* 14.4 (July 2015), pp. 689–698. https://doi.org/10.1109/TNANO.2015.2429893.

[Tro+16] J. Trommer, A. Heinzig, T. Baldauf, T. Mikolajick, W. M. Weber, M. Raitza, and M. Völp. "Reconfigurable nanowire transistors with multiple independent gates for efficient and programmable combinational circuits". In: *2016 Design, Automation Test in Europe Conference Exhibition (DATE)*. Mar. 2016, pp. 169–174.

[WH15] Neil HE Weste and David Harris. *CMOS VLSI design: a circuits and systems perspective*. Pearson Education India, 2015.

[Zha+14] Jian Zhang, Xifan Tang, Pierre Emmanuel Gaillardon, and Giovanni De Micheli. "Configurable Circuits Featuring Dual-Threshold-Voltage Design With Three-Independent-Gate Silicon Nanowire FETs". In: *IEEE Transactions on Circuits and Systems I: Regular Papers* 61.10 (Oct. 2014), pp. 2851–2861. https://doi.org/10.1109/TCSI.2014.2333675.

Chapter 5
Logic Synthesis with XOR-Majority Graphs

In the previous chapter, it was shown that RFETs allow efficient implementation of *self-dual* logic functions with fewer transistors. The Boolean property of self-duality was demonstrated to be the natural abstraction for reconfigurable nanotechnologies. The availability of self-dual gates in the generic library also leads to better area results during technology mapping. This chapter takes a step further in the synthesis flow and explores logic representations that can natively utilize the self-dual property during logic optimization to get further improvements. This chapter focuses on improving the logic synthesis and the technology mapping flow to achieve better area results for circuits based on emerging reconfigurable nanotechnologies by preserving the self-duality of a given circuit.

Logic synthesis plays an important role in optimizing a logic representation for a given circuit in terms of a cost function, typically focusing on the reduction of area or delay. At the technology-independent level, multilevel logic representations are used to represent and optimize circuits. Recently, novel representations have been proposed that enhance *And-inverter graphs* (AIGs) [KGP01] and *Majority-inverter graphs* (MIGs) [AGM16] with an additional XOR primitive. These new logic representations, called *Xor-And graphs* (XAGs) [HFS17] and *Xor-Majority graphs* (XMGs) [Haa+17, Chu+19], offer better compaction and, thus, often have a positive effect on the performance of logic minimization techniques [Haa+17]. Technology mapping, on the other hand, focuses on expressing the minimized logic representation in terms of a network of logic gates chosen from a given library [Cha07].

XMG-based synthesis flow is explored for reconfigurable nanotechnologies to preserve and utilize the existing self-duality of a circuit, as they are less prone to disrupting self-duality as compared to AIGs. XMG's gate primitives—2-input XORs and Majority offer a compact representation to abstract self-dual logics because Majority and odd-input XORs are self-dual.

This chapter provides experimental evidence to demonstrate that a logic representation that compactly abstracts self-duality can preserve more self-duality in the

S. Rai, A. Kumar, *Design Automation and Applications for Emerging Reconfigurable Nanotechnologies*, https://doi.org/10.1007/978-3-031-37924-6_5

circuit and achieve better area results for RFET-based circuits. This is motivated by the fact that synthesis approaches such as logic optimization and technology mapping involve structural changes over a graph representation of a circuit which can disrupt the existing self-duality of a circuit. Experiments show that for circuits with a high percentage of self-dual logic (e.g., Majority and odd-input XORs), AIGs are more prone to decompose self-dual logic.

Research Question As CMOS favors negative unate Boolean logic, AIGs are the natural abstraction for CMOS logic [Ama+15]. However, are AIGs appropriate for RFET-based circuits as well? This chapter investigates this question.

5.1 Contributions

The major contributions of this chapter are as follows:

- To preserve self-duality, an XMG-based synthesis flow is presented for circuits based on reconfigurable nanotechnologies. Experimental evaluation demonstrates that the XMG-based synthesis flow enables a better area reduction for RFET-based circuits.
- A resubstitution and a rewriting algorithm for XMGs are proposed. The two techniques play an important role in XMG optimization and also increase the self-duality density of the circuits.
- The resubstitution algorithm uses a new filtering rule for 3-input XOR gates (XOR3) that drastically reduces the runtime.
- Using an area-oriented mapping, it is demonstrated that XMGs enable a higher percentage of self-dual cuts during mapping compared to AIGs. The higher share of self-dual cuts subsequently leads to more mapping opportunities on to self-dual logic gates.

The proposed XMG-based approach is compared with three different flows. First, it is compared with the native flow of the state-of-the-art tool ABC [BM10] (baseline). Second, XMG-based logic synthesis is used as a starting point and the area results calculated by the recently proposed logic representation agnostic mapper [Tem+21] are compared with those calculated by the ABC technology mapper. Third, since the logic representation agnostic mapper can support different logic representations, the proposed XMG-based approach is compared with the AIG-based approach within the *mockturtle* framework [Soe+18].

The experiments are performed over two sets of benchmarks to evaluate the approach. In the first set of experiments, synthetic benchmarks are enumerated with varying degrees of self-duality to show that the XMG-based approach gives the best results across these three different flows. In the second set of experiments, cryptographic benchmarks [Alb+15, Cha+17] are used. First, it is demonstrated that the XOR3 filtering rule leads to 59.48% improvement in runtime during XMG-based resubstitution. In terms of improving the self-duality of a given circuit, the two proposed techniques—resubstitution and rewriting in conjunction—lead to a 23% average increment of self-duality over the cryptographic benchmark suite. Second, the proposed XMG-based approach leads to an area savings of up to 12.3% with respect to the baseline. Last, the relation between self-dual cuts and the final share of the area by self-dual logic gates as a percentage of the overall area is explored. On average, the XMG-based approach results in 8.68% more area occupied by self-dual gates compared to the AIG-based approach within the *mockturtle* framework. These comparisons show that for circuits with higher self-duality, the proposed XMG-based logic synthesis in conjunction with the proposed mapper achieves superior results compared to other contemporary flows.

5.2 Organization

The chapter is organized as follows: Sect. 5.3 presents a motivating example of a simple circuit and compares the AIG and the XMG flow. Section 5.4 introduces the necessary preliminaries and discusses earlier works based on XMGs. Section 5.5 explains how self-duality can be preserved using XMG-based logic synthesis and technology mapping. This is followed by Sect. 5.6, which explains the proposed resubstitution and the rewriting algorithm for XMGs. Then, in Sect. 5.7, the versatile mapper is explained. This is followed by Sect. 5.8, which presents details about the algorithm to generate benchmarks with varying degrees of self-duality. Section 5.9 contains details about experimental evaluations. Closing remarks and future research directions can be found in Sect. 5.10.

5.3 Motivation

In order to understand the motivation behind this work, let us consider a circuit that consists of a single 3-input XOR. Our objective is to carry out technology mapping of this given circuit. Two straightforward mappings for the given circuit are possible—one where the circuit is mapped to two 2-inputs XOR gates and the other directly to a 3-input XOR gate. From a CMOS perspective, the mapper should prefer the first mapping as we know that a 3-input XOR logic is avoided in CMOS as it requires multiple transistors and often has large delay [Rai+19]. Logic gates with many CMOS transistors require cascading or branching of multiple transistors that hampers the performance of such logic gates based on CMOS. Hence, circuits based on CMOS prefer logic gates with few inputs as it is better in terms of performance and area. This is also one of the considerations in contemporary logic representation of AIG as CMOS favors negative unate logic [Ama17]. However, in the case of RFETs, a 3-input XOR is a preferred mapping since it is a self-dual logic gate [Rai+20]. From a logic synthesis perspective, the contemporary logic representation of AIG uses 6 nodes and 13 edges to represent the circuit. In contrast,

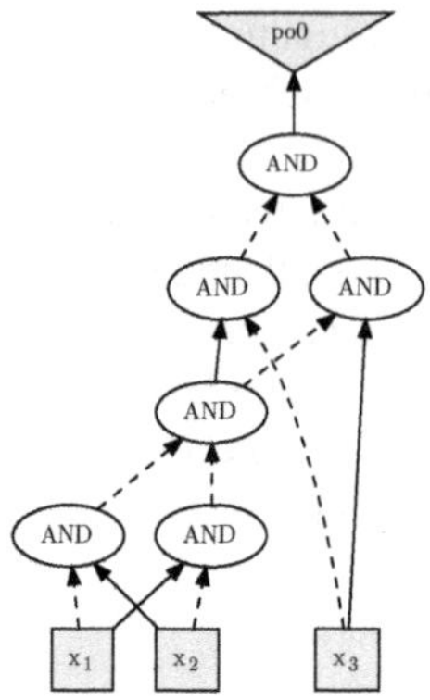

(a)

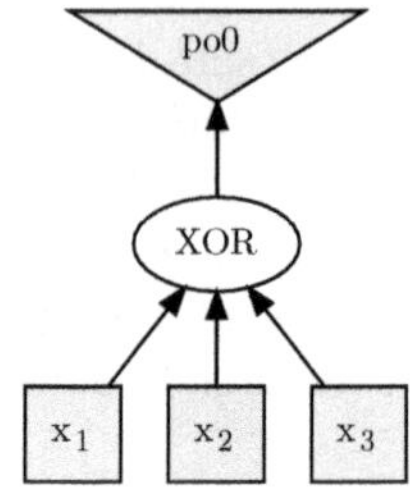

6

(b)

Fig. 5.1 Edge and node representation in case of AIG and XMG representation for the function, $f = x_1 \oplus x_2 \oplus x_3$ (3-input XOR). The AIG has 6 gates and 13 edges, while the XMG has 1 gate and 4 edges. (**a**) AIG. (**b**) XMG

XMG uses 1 nodes and 4 edges. Both the AIG and XMG representations are shown in Fig. 5.1. If we consider the AIG representation, a simple 3-input cut-based technique during logic optimization and technology mapping results in the first kind of mapping. However, in the case of XMGs, with the same setting, the second mapping is achieved. The higher number of edges and nodes in AIG compared to XMG explains this difference in mapping results. This simple intuitive example motivates us to explore XMGs for RFET-based circuit. As CMOS favors negative unate logic gates, AIGs are the natural abstraction for CMOS logic [Ama17]. However, are AIGs appropriate for RFET-based circuits as well?

5.4 Background and Preliminaries

In the previous chapter, it was shown that self-dual functions are a logical abstraction for ambipolar nanotechnologies. The multiple functionalities exhibited by RFET-based logic gates are due to the switchable pull-up and pull-down networks, as shown in Fig. 4.1b. The switching of polarities of individual transistors in their respective pull-up and pull-down networks is caused by the change of the potential at the program gate, as shown in Fig. 4.2a and b. This change of the potential causes the PFET to become an NFET and vice versa, which in-effect switches the pull-up and pull-down networks, as shown in Fig. 4.1b. This corresponding switch in electrical behavior is abstracted conveniently by a self-dual Boolean function. Only with self-dual functions, the polarity switch in individual transistors creates a conducting path between the output and the source (or drain) leading to the realization of the dual of the original function.

5.4.1 Terminologies

Terminologies that are used throughout the rest of the chapter are introduced here. A given circuit is represented as a directed acyclic graph (DAG) consisting of nodes and edges. Nodes are data structures representing logic gates as defined by a given logic representation (AIG, XMG, etc.). Edges denote the connections between nodes. Without losing generality, the terms *circuit*, *logic graph*, and *logic network* are used interchangeably throughout this chapter. Next, we look at two terminologies which will be used in the experiment Sect. 5.9 of this chapter.

1. *Self-duality density*: *Self-duality density* for a circuit (or a logic network) is defined as the ratio of the total number of self-dual nodes to the total number of nodes.
2. *Cut-enumeration techniques*: In order to apply strong Boolean optimization algorithms, k-feasible cuts per node are needed to be computed. In the subsequent sections, we see that both rewriting approach and technology mapping algorithm utilize methods involving k-feasible cuts per node more commonly known as *cut-enumeration* techniques.

5.4.2 Self-Duality in NPN Classes

As stated in Theorem 4.1, self-dual functions are rare. Table 5.1 shows the distribution of the self-dual functions over all Boolean functions up to 4 variables and their NPN representatives. NPN canonization preserves the self-duality of a Boolean function, i.e., if a Boolean function is self-dual, so are all Boolean functions obtained by applying the NPN transformations to it. The numbers in the table illustrate that self-dual functions are not only rare when compared to the total number of Boolean functions, but also show that they reduce with increasing number

Table 5.1 Distribution of self-dual functions in NPN

	Functions		NPN classes	
No of vars.	Self-dual + norm.	Total	Self-dual + norm.	Total
1	$1+3$	4	$1+1$	2
2	$4+12$	16	$1+3$	4
3	$16+240$	256	$3+11$	14
4	256 + 65,280	65,536	$7+215$	222

of variables. Whereas 25% of the NPN representatives in 2 variables are self-dual, this percentage drops to 21.43% and 3.15% for 3 and 4 variables, respectively.

5.4.3 Majority Logic Synthesis

In this section, we review majority logic synthesis as explained in [Ama+16]. The majority function $\langle x_1, x_2, x_3 \rangle$ of three Boolean variables x_1, x_2, x_3 evaluates to true if and only if at least two of the three variables have a value of true. The majority function can be expressed in disjunctive normal form or conjunctive normal form, i.e.,

$$\begin{aligned}\langle x_1, x_2, x_3 \rangle &= x_1x_2 + x_1x_3 + x_2x_3 \\ &= (x_1 + x_2)(x_1 + x_3)(x_2 + x_3). \end{aligned} \tag{5.1}$$

By setting one of the three Boolean variables x_1, x_2, x_3 in the majority function $\langle x_1, x_2, x_3 \rangle$ to a constant value 0 or 1, one obtains the logic functions AND and OR, respectively, i.e.,

$$\langle 0, x_1, x_2 \rangle = x_1x_2 \text{ and } \langle 1, x_1, x_2 \rangle = x_1 + x_2. \tag{5.2}$$

Vergelijking 5.2 is often called the *containment rule* of majority.

5.4.4 Earlier Work on XMG

The authors of [Haa+17] proposed a logic representation, called *XOR-MAJ graph* (XMG), consisting of three-fan-in majority (MAJ) gates, three-fan-in XOR gates, and inversion. The representation enables a size-proportional representation of both, n-input unate and n-input binate logic functions. XMGs were first introduced in [Haa+17] as a means for the underlying logic representation for exact synthesis. Exact synthesis solves the problem of finding an optimum network for a given function. Since exact synthesis uses SAT solving or enumeration, the runtime of exact synthesis tools directly depends on the size of the logic representation. Since XMGs have both binate (*XOR*) and unate (*MAJ*) nodes, this results in a size-proportional logic representation for both n-input unate and n-input binate functions compared to the representation of binate logic (*XOR*-based logic) of MIGs or AIGs [Chu+19]. Algebraic optimizations for XMG-based logic synthesis were proposed in [Chu+19]. The authors explored Boolean algebraic optimizations for *XOR* and *XOR-MAJ* logic and were able to achieve depth optimizations.

5.4.5 *Classification of Boolean Functions*

Two Boolean functions $f(x)$ and $g(x)$ over the variables $x = x_1, \ldots, x_n$ belong to the same class C of functions if they are equivalent modulo some fixed set T of function transformations. In other words, if f can be transformed into g (or vice versa) by applying a sequence of Boolean transformations from T, then f and g are T-equivalent. The three most common function classes are (1) P, permutation of inputs; (2) NP, negation of inputs and permutation of inputs; and (3) NPN, negation of inputs, permutation of inputs, and negation of outputs. The three function classes are related as follows: $P \subseteq NP \subseteq NPN$. These classes play an important role in technology-independent logic synthesis since two functions belonging to the same class can be represented with the same graph structure modulo the respective input and output transformations [Hua+13]. For example, the functions $f_1 = x_1x_2 + x_3$ and $f_2 = x_1 + x_2x_3$ are NPN-equivalent because by swapping the variables x_1 and x_3 in f_1, the function f_2 is obtained. Hence, if a node-minimum AIG for f_1 is known, then a node-minimum AIG for f_2 can be derived by swapping the inputs x_1 and x_3.

In the following, function classification (or function canonization) is used to perform Boolean matching in logic synthesis and technology mapping techniques. Boolean rewriting requires a database of size-minimum circuits for all Boolean functions. With the help of function classification [MB06], the database can be reduced to one size-minimum circuit per class. In technology mapping, the pre-enumeration and hashing of the NP-equivalent functions of all cells in the technology library enable us to match Boolean functions with cells more quickly [Cha07].

5.5 Preserving Self-Duality

5.5.1 *During Logic Synthesis*

Due to their reconfigurability at the device level, RFETs enable efficient implementations of nontrivial self-dual logic functions in terms of the number of transistors [Rai+20] compared to CMOS. For example, a XOR logic gate with 3 inputs (shown in Fig. 4.2b) requires $4 + 2$ (for P and P$'$) transistors when using RFETs as compared to 22 transistors when realized in CMOS technology [Rai+19]. This implies that circuit implementations with RFETs lead to area reductions if they have a high density of nontrivial (3 or more input functions) self-dual gates. In order to use this property, it is therefore imperative that the self-duality in a logic representation is preserved through logic optimizations. From a logic representation perspective, if we consider AIGs (consisting of two-input AND gates with complement-edge attributes), a nontrivial self-dual function is decomposed into multiple AIG nodes. Similarly, for MIGs, parity-based self-dual functions (odd

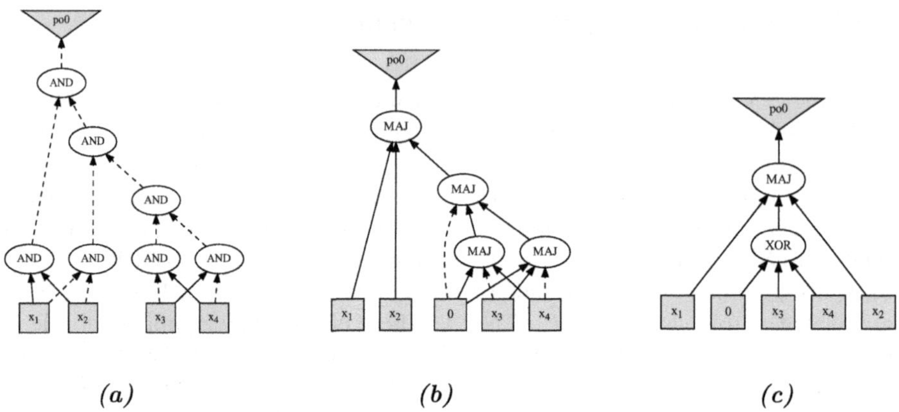

Fig. 5.2 Different logic representations of the function, $f = \langle x_1, x_2, (x_3 \oplus x_4)\rangle$. One can notice that the number of nodes and edges are the lowest in the XMG representation. (**a**) AIG. (**b**) MIG. (**c**) XMG

input XOR logic) cannot be represented in a compact manner using *MAJ* nodes alone [Chu+19]. Various logic optimization techniques using cut-based techniques on AIGs and MIGs can allocate different cuts for the decomposed self-dual logic, thereby losing self-duality.

In contrast, XMGs use *XOR* and *MAJ* nodes as logic primitives. Each MAJ and odd-input XOR function is self-dual, so using XMGs can better preserve self-duality during logic optimization compared to other logic representations. This can easily be seen in Fig. 5.2. The figure shows three logic representations of the same function $f = \langle x_1, x_2, (x_3 \oplus x_4)\rangle$. The AIG logic representation requires 7 gates, while the same function has 4 and 2 gates when represented as MIG and XMG, respectively. In the example, there are more edges in the AIG and MIG representations than in the XMG representation. This leads to an increase in the number of competing structural cuts of the logic network in logic optimization and technology mapping phase. With the above benefits in mind, we have developed an XMG-based logic optimization flow that addresses these issues and helps to achieve area reductions for RFET-based circuits.

5.5.2 During Versatile Technology Mapping

One of the limitations of the previous work [Rai+21] is the absence of a technology mapper that can map with arbitrary logic representations. The authors in [Rai+21] used XMG as the graph representation and carried out logic optimization intending to preserve self-duality. However, technology mapping was performed using ABC's native technology mapper, where AIG is the default logic representation. Thus, the XMG graph representation has to be converted to an AIG graph representation.

Algorithm 5.1 Boolean filtering and resubstitution

Data: Window W in a logic network with root node n
Result: Node resubstitute for n or $\perp$ if no resubstitution has been found

```
Set M ← W.computeMFFC(n)  Set D ← W.collectDivisors(n)\M  Set TT ← W.simulate()
sortByDBP(D, TT, n)  for i ← 0 to |D| do
    if 3 · DBP(D[i]) < DBP(n) then
        return ⊥
    end
    for j ← i + 1 to |D| do
        if DBP(D[i]) + 2 · DBP(D[j]) < DBP(n) then
            break
        end
        for k ← j + 1 to |D| do
            if f = TT[i] ⊕ TT[j] ⊕ TT[k] then
                return W.xor3_resub(n, D[i], D[j], D[k])
            end
            if f = ¬TT[i] ⊕ TT[j] ⊕ TT[k] then
                return W.xor3_resub(n, ¬D[i], D[j], D[k])
            end
        end
    end
end
return ⊥
```

During this process, individual XMG nodes are represented with multiple AIG nodes. This conversion leads to an increase in the number of competing cuts for large self-dual logic nodes. This can be understood from Fig. 5.2 where the self-dual nodes of MAJ are represented using multiple AIG nodes. This increase in the number of competing cuts during mapping can disrupt the self-duality density of the circuit leading to suboptimal results in the context of area reduction for RFET-based circuits. In our experiments, we found that in comparison to the native AIG-based technology mapping in ABC, this XMG-based logic graph leads to a tripling of the number of competing cuts during technology mapping within ABC. As a result, several optimal cuts that can preserve self-duality are lost during the technology mapping phase. Therefore, a logic-representative agnostic technology mapping is essential for our work. We have explored this observation and then explained it in Sect. 5.9.4.

5.6 Advanced Logic Synthesis Techniques

While the prior works [Haa+17, Chu+19] introduced XMGs and algebraic optimizations for them, the repertoire of Boolean methods with a resubstitution and NPN-based cut-rewriting technique is extended.

5.6.1 XMG Resubstitution

Boolean resubstitution is a logic optimization method that re-expresses the function of a node n in a logic network N using nodes, called *divisors*, already present in N. Nodes that are exclusively used by n and are not required by any other logic in the logic network become free and can be removed. A resubstitution leads to a size reduction if the number k of newly added nodes to re-express a node's function is less than the number l of removed nodes in its *maximum fanout-free cone* (MFFC, [MB06]).

Resubstitution algorithms are available for different multilevel logic representations including AIGs [MCB06, MB06], MIGs [Rie+18, Rie+19b], and logic networks [KK04, Mis+11, Ama+18] focusing on two-input AND operations, three-input MAJ operations, and combinations of two-input gates such as XOR-ANDs, AND-XORs, or three and two-input gates such as MUX-XORs, respectively.

Computing three-input XOR (3-input XOR is a self-dual logic gate), resubstitutions are particularly time-consuming because divisor filtering techniques developed for AND and OR operations cannot be applied. To substitute a node n in a network with logic function $f_n(x)$ by a three-input XOR operation, three divisor nodes d_1, d_2, and d_3 have to be found, such that

$$f_n(x) = f_{d_1}(x) \oplus f_{d_2}(x) \oplus f_{d_3}(x) \tag{5.3}$$

for all assignments to the primary inputs x, where f_{d_1}, f_{d_2}, f_{d_3} are the divisor functions, respectively.

State-of-the-art Boolean resubstitution algorithms over-approximate the node functions using windowing to apply scalable truth-table computations. The algorithms have to iterate over all triples of nodes in a window of a root node n (excluding the root node's MFFC) to test if Eq. (5.3) holds. The first substitution possible that reduces the network's size is accepted. In the worst case, if no resubstitution can be accepted, $\mathcal{O}(w^3)$ tests are required for a window with w nodes.

Filtering techniques help to reduce the number of tests required and significantly speed up the performance of resubstitution algorithms in practice. A new filtering rule is developed for three-input XORs guiding the search for divisors using distinguishing bit pairs [CMB08]: a resubstitution of a target node n with function $f(x)$ and divisor nodes d_1, d_2, d_3 with functions $f_{d_1}(x)$, $f_{d_2}(x)$, $f_{d_3}(x)$ over common window inputs x exists if and only if for any pair $\hat{x}_i \neq \hat{x}_j$ of input assignments

$$f(\hat{x}_i) \neq f(\hat{x}_j) \implies \bigvee_{1 \leq a,b \leq 3, a \neq b} d_a(\hat{x}_i) \neq d_b(\hat{x}_j). \tag{5.4}$$

Example 5.6.1 Suppose that n is a node to be substituted and $D = \{d_1, d_2, d_3, d_4\}$ are divisors with the following truth tables:

Node	TT	DBP
n	`1001 0110`	16
d_1	`0000 0100`	4
d_2	`0111 1111`	4
d_3	`1000 0000`	4
d_4	`1111 1011`	4

The absolute distinguishing bit power $\text{DBP}(n) = 16$, whereas the relative distinguishing bit powers $\text{DBP}_n(d_i) = 4$ for $i \in \{1, \ldots, 4\}$. We use a counting argument as a necessary condition to conclude that no Boolean operation using three (or less) of the given divisor functions is sufficient to synthesize n. Assuming that the given divisor functions distinguish different bit pairs, any subset of D of size 3 can distinguish at most 12 bit pairs. However, n requires 16 bit pairs to be distinguished. In other words, regardless of which three divisors one picks, there is always at least one bit pair that cannot be distinguished. This can be easily verified by looking at the truth tables of the divisor functions: the two least significant bits of all divisors functions are equal, but the two least significant bits of n are not.

Utilizing Eq. (5.4), all divisor nodes are sorted in a window by the number of bit pairs distinguished by the divisor with respect to the root node's target function. The *absolute distinguishing bit power* $\text{DBP}(n)$ of the root node n is defined as the number of pairs $(\hat{x}_i, \hat{x}_j)$ of input assignments for which $f_n(x_i) \neq f_n(x_j)$. Similarly, the *relative distinguishing bit power* $\text{DBP}_n(d)$ of a divisor d is defined as the number of pairs $(\hat{x}_i, \hat{x}_j)$ of input assignments for which $f_n(\hat{x}_i) \neq f_n(\hat{x}_j)$ and $f_d(\hat{x}_i) \neq f_d(\hat{x}_j)$.

Algorithm 5.1 shows the Boolean filtering and resubstitution algorithm as pseudo-code. The divisors are sorted (line 4) by their relative distinguishing bit power—a higher relative distinguishing bit power will more likely lead to a possible resubstitution. The relative distinguishing bit power is further leveraged to filter *insufficient* divisor triples. Given a sorted list $D = d_1, \ldots, d_w$ of divisors such that $\text{DBP}_n(d_i) \geq \text{DBP}_n(d_j)$ for all $i < j$, a single divisor d can never be completed to a divisor triple that passes the test in Eq. (5.3) if $3 \cdot \text{DBP}_n(d) < \text{DBP}(n)$ (line 6). Since the list is sorted, no remaining divisor will pass this test either, such that the algorithm can terminate (line 7). For a similar reason, no divisor pair d_i, d_j, $i < j$, can be completed to a divisor triple that passes the test in Eq. (5.3) if $\text{DBP}_n(d_i)+2\cdot\text{DBP}_n(d_j) < \text{DBP}(n)$ (line 9). In this case, the algorithm can proceed by selecting another candidate divisor d_i (line 10).

5.6.2 Exact XMG Rewriting

Boolean rewriting is a logic optimization method that selects small parts of a logic network and replaces them with more compact implementations to reduce its number of nodes. State-of-the-art rewriting algorithms either rely on a database of precomputed size-optimum subnetworks for all Boolean functions up to 5 inputs [MB06] or compute size-optimum subnetworks on the fly using exact synthesis [Rie+19a, RMS20]. DAG-aware rewriting [MB06], fast cut enumeration techniques [CWD99a], NPN canonization [Hua+13] of Boolean functions, and efficient caching [RMS20] enable scalability.

Rewriting XMGs has been first proposed in [Haa+17] using a two-step approach: (1) A logic network is mapped into a network of k-feasible lookup tables (LUTs); (2) the k-feasible LUTs are resynthesized into size-optimum XMGs. By repeating the two steps until convergence, substantial size reduction can be achieved.

An improved XMG rewriting approach, called *exact XMG rewriting*, is proposed that integrates both steps into a single algorithm. For each node, in the logic network, the set of all k-feasible cuts is enumerated, each cut is simulated to obtain its Boolean functions, and the functions are resynthesized using exact synthesis. In contrast to the previous approach [Haa+17], the proposed algorithm takes advantage of structural hashing to utilize the existing logic within the network, such that a global size reduction can be achieved even if a locally smaller subnetwork is replaced with a larger subnetwork.

The algorithm can be parameterized with a set of gate primitives and supports synthesis of multiple candidates per cut function. A conflict limit in exact synthesis allows us to limit the maximum synthesis effort per function. An exact XMG rewriting for three different sets of gate primitives is considered:

1. Three-input MAJ gates with two-input XOR gates as originally proposed by [Haa+17].
2. Three-input MAJ gates and three-input XOR gates to enable a more compact representation of Boolean functions. Note that with constants the three-input XOR gate can simulate the function of two-input XOR gates and, thus, is a generalization of two-input XOR.
3. Three-input MAJ gates without constants and three-input XOR gates to improve the internal self-duality of a logic network during rewriting.

In practice, when mapped to RFETs, the best optimization in terms of area is achieved with the first set of gate primitives. The MAJ gate and the three-input XOR gates are large primitives, and hence, fine granular optimizations can be lost. This is an inverse scenario that is explained in Sect. 5.5.2. Since the individual nodes are large, certain competing cuts (which could have been generated using smaller logic representation) required for optimal area reduction are lost. Also, for circuits, which do not have high self-duality density, the optimizations are suboptimal as the obtained graph representation has a much higher number of edges (due to the undue presence of constants) than that possible with smaller gate primitives. The increase in these constant edges leads to poor area optimizations.

5.7 Logic Representation-Agnostic Mapping

This section discusses the details of the versatile mapper used for mapping natively on XMG-based logic representations.

5.7.1 Versatile Mapper

In design automation, mapping is the process of expressing a logic graph using a set of primitives. Commonly, mapping either refers to a LUT mapping or a standard-cell mapping [Mic94]. In this chapter, standard-cell mapping in the context of emerging nanotechnologies is considered. Once all the logic optimizations are carried out over a logic graph during the logic synthesis phase, the logic graph is converted to a k-bounded network, called subject graph. During technology mapping, the subject graph needs to be expressed using the standard cells present in a given cell library. This matching is often carried through Boolean matching [MM90].

The versatile mapper supports arbitrary graph representations, such as AIG, XMG, XAG, or MIG, to represent the subject graph in standard cells (as given in a technology library). In this work, the technology library is an RFET-centric generic library consisting of logic gates as proposed in [Rai+19]. The mapping can be optimized for either area or delay. The mapper follows four main steps:

1. *Library generation:* In the library generation phase, the mapper generates a hash table for the gate primitives listed in the technology library (`.genlib`). The hash table consists of NP enumerations for each of the gate primitives subject to filtering of enumerations that are functionally symmetric [Mic94]. A data structure is maintained that contains the delay and area values of each gate configuration.
2. *Cut enumeration:* In this step, the logic network is traversed in topological order and cuts are enumerated with up to k inputs. For each of the cuts, truth tables are computed which are later used during Boolean matching to find a match in the hash table of the gate primitives. The cut enumeration technique used in the mapper is based on priority cuts as proposed in [CWD99b, Mis+07].
3. *Boolean matching:* In this step, the truth table for each cut of the subject graph is looked up in the hash table (generated in the first step) to select gates that can implement it. Both polarities of the cut are considered during mapping to enable logic sharing of inverters and to avoid additional inverter delay.
4. *Optimization objective:* Once the matching is done for each of the cuts, a *cover* is selected. A *cover* is a set of cuts so that all cuts in the set are either rooted at the primary output or at the leaves of another cut. The cover is selected so that an optimum area or delay for the circuit is achieved. During delay minimization, the primary objective is to have the smallest delay of the largest path of the cover, and during area minimization, the area of the cover is minimized. Various

heuristics [MBV06, CWD99b] are followed during this step to attain the best possible mapping.

For more details on actual implementation of the mapper, readers are requested to refer to [Tem+21].

5.7.2 Support of Supergates

Using supergates is an efficient technique as suggested by Chatterjee et al. to mitigate structural bias [Cha+05]. Structural bias arises from the fact that the structure of the starting logic graph representation dictates the final mapping quality to a large extent. By combining several gate primitives from the cell library, a list of supergates is precomputed to be used later during the technology-independent mapping step. Supergates aim to explore unique combinations of gate primitives which otherwise cannot be used during technology mapping [Cha+05]. An example is shown in Fig. 5.3. Here, *MUX* is the root gate, and its two inputs are connected to outputs of two other logic gates. Consequently, a 5-input supergate is realized. Similarly, various other combinational supergates are precomputed and added to the hash table so that they are available for matching during technology-independent mapping. From the previous section, an obvious outcome is the increase in the size of the hashtable created during the library generation stage. Supergates lead to improved quality of mapping at the expense of requiring additional runtime [Cha07].

The implementation can read supergate libraries produced by the open-source tool ABC [BM10]. For each entry in a `.super` file, the implementation computes the truth table, the area, and the delay. It is then added to the hash table generated in the library generation step. Once added to the hash table, the supergates are available to the mapper during the Boolean matching step.

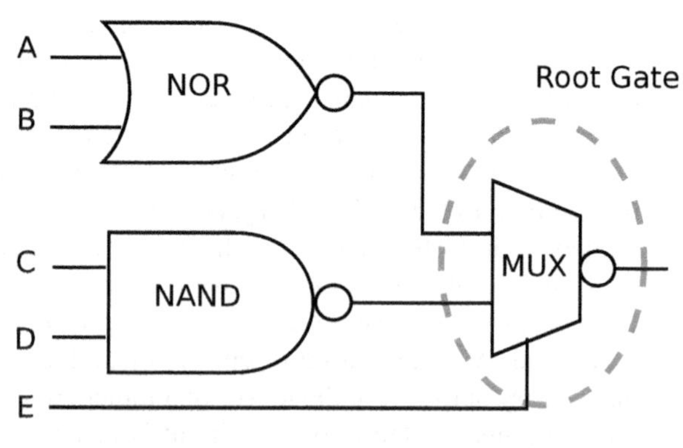

Fig. 5.3 Supergate generation. Here, MUX is the root gate with NANDs and NORs as the new input pins. The supergate thus generated has five inputs

5.8 Creating Self-Dual Benchmarks

To evaluate the efficacy of the approach compared to state-of-the-art logic synthesis approaches for RFET-based standard-cell mapping, synthetic benchmark circuits are generated with varying numbers of self-dual logic gates. A simple graph-based technique is adopted to generate benchmark circuits with varying numbers of self-dual logic gates. These benchmarks are built in a level-by-level fashion from the primary inputs to primary outputs. There have been multiple previous works targeting benchmark generation [Net+19, SVC00].

Algorithm 5.2 is used to generate benchmarks with different self-dual densities, starting from an empty XMG network. The algorithm takes the following four parameters as inputs: *the number of primary inputs (PIs)* (num_pis), *the number of levels* ($levels$), *the number of nodes per level* ($nodes\_per\_level$), and *the self-duality index* ($index$). For a given set of the above three parameters—num_pis, $levels$, $nodes\_per\_level$—10 different benchmarks are generated by assigning values $\{1 \rightarrow 10\}$ sequentially to the self-duality index ($index$). A self-duality index value v implies that out of every 10 nodes added, v are self-dual nodes (MAJ or 3-input XOR) and $10 - v$ are normal nodes (AND, OR, 2-input XOR) (line 8–11). For example, let us consider where the values of the parameter given to Algorithm 5.2 as $num\_pis = 12$, $levels = 512$, $nodes\_per\_level = 131$. In this case, Algorithm 5.2 creates 10 different circuits, all with 12 primary inputs, 512 levels between the primary inputs and primary outputs, and 131 maximum nodes in each level. Each benchmark is guided with the self-duality index value ($index$) chosen sequentially from $\{1 \rightarrow 10\}$ so that the 10 benchmarks have varying levels of self-duality density. The algorithm maintains a list of signals to keep track of

Algorithm 5.2 Generate self-dual XMG network

```
Data: num_pis, levels, nodes_per_level, index
Result: XMG network N
Set signalList ← [] Set counter ← 0 for k ← 0 to num_pis do
  | signalList.add(N.create_pi())
end
for i ← 0 to levels do
    for j ← 0 to nodes_per_levels do
        fanins ← signalList.randSubSet() if counter < index then
          | node ← N.create_selfdual_gate(fanins)
        else
          | node ← N.create_normal_gate(fanins)
        end
        signalList.add(node) counter ← (counter + 1) mod 10
    end
end
for o ∈ signalList.not_used() do
  | N.create_po(o)
end
return N
```

all generated nodes (line 12). The algorithm first generates the primary inputs of the XMG network and adds them to the list (line 3–4). It then adds new gates in a level-by-level fashion by randomly selecting fan-ins from the updated signal list (line 7). It is to be noted that the self-duality index value of v (let us say 5) does not correspond to $(10 \times v)\%$ (50%) of the self-duality density. This is primarily because during the construction of the circuit, nodes are added only after checking whether another node with the same fan-ins already exists in the graph or not. In this way, XMG optimizes away some of the redundant nodes. Finally, those nodes that are never referenced by any other node are marked as primary outputs (line 14–15). The source code of the benchmark generator is available online.[1]

5.9 Experiments

In this section, the experimental setup is described and the obtained results are discussed. All algorithms have been implemented in the open-source logic synthesis tool *mockturtle* from the EPFL logic synthesis libraries [Soe+18]. For technology mapping, the RFET-based generic library consisting of logic gates as mentioned in [Rai+19] is used.

5.9.1 XMG-Based Flow

The proposed XMG-based flow comprises of a logic synthesis and a technology mapping flow. XMGs are used as the graph representation for a given circuit throughout these two steps. The following steps are carried out in this flow:

1. Starting with the graph representation of a given circuit (let us say AIG), it is first converted into a 4-input LUT-based logic graph. The LUT-based representation thus obtained is then resynthesized,[2] where each 4-input LUT of the LUT graph is converted into an equivalent XMG using a pre-generated 4-input NPN class for XMGs.
2. The proposed two algorithms (resubstitution and 4-input NPN-based rewriting) are then applied iteratively to the obtained XMG until no size improvement of more than 5% is possible
3. Finally, the versatile technology mapper is used with the obtained XMG from step 2. The mapper uses the RFET-based generic library and computes the supergates to achieve an area-oriented mapping.

[1] https://github.com/shubhamrai26/self-dual-experiments.

[2] Calling *node resynthesis* technique as defined in mockturtle framework [Soe+18].

5.9.2 Experimental Setup

Experiments are carried out on two sets of benchmarks. First, the synthetic benchmarks are generated (as described in Sect. 5.8) and the post-mapping areas for different flows (mentioned below) are compared. This gives an empirical evidence of the proposed XMG-based synthesis flows. Then, the proposed approach is evaluated using real benchmarks in the form of cryptographic benchmarks [Alb+15, Cha+17]. For post-mapping area, the following synthesis flows are used:

1. **abc_rwrs:** Native ABC flow is used to carry out both logic optimization and technology mapping. Here, AIG is the used graph representation. This flow is the *baseline* in the experiments.
2. **mtl_AIG:** AIG is the used graph representation. The versatile technology mapper is used within the mockturtle framework to compute post-mapping area.
3. **abc_XMG:** The flow as mentioned in [Rai+21] is used. Here, *mockturtle* is used for XMG-based logic optimizations. Technology mapping is, however, carried out with ABC.
4. **Proposed:** This is the proposed XMG-based flow.

Then, the proposed approach is evaluated using real benchmarks in the form of cryptographic benchmarks [Alb+15, Cha+17]. A detailed analysis is carried out for our individual contributions. The area comparison is carried out in terms of number of transistors. Since we are comparing different logic synthesis flows, a reduction in the number of transistors has a direct impact on reducing the overall area of the RFET-based chip. This is the normalized area as mentioned in the `.genlib`. The three flows (mtl_AIG, abc_XMG, and ours) are compared with the baseline flow in terms of percentage. A negative percentage implies that the baseline flow is better than the other flows, while positive percentages imply these flows are better than the baseline.

Throughout the experiments, for the flows that use XMGs, the *self-duality density* is calculated as follows:

$$\text{sd-density} = \frac{\text{number of XOR3 nodes} + \text{number of MAJ nodes}}{\text{total number of nodes}} \tag{5.5}$$

The metric is used to indicate the amount of self-duality that exists in the graph representation of the circuit.

5.9.3 Synthetic Self-Dual Benchmarks

Algorithm 5.2 is used to generate 100 sets of benchmarks. For each set, the self-duality index in Algorithm 5.2 is iterated from 1 to 10 to generate 10 benchmarks for a particular set of parameters. Hence, in total, there are 1000 benchmarks to

Table 5.2 Post-mapping area comparison for synthetic self-dual benchmarks using different synthesis flows. It can be noticed that the proposed XMG-based flow gives the best results. Additionally, it can be noted that the gain (%) is higher for circuit with higher self-duality density. Hence, improvement increases with higher SD index

SD index	SD density	Baseline abc_rwrs	mtl_AIG (%)	abc_XMG (%)	Proposed (%)
1	6.63	29904.22	0.93	−0.13	1.15
2	14.97	34248.61	0.79	−0.06	1.07
3	14.97	37732.53	0.62	−0.07	0.92
4	29.61	41296.80	0.57	−0.02	0.94
5	29.85	44717.89	0.62	0.05	1.03
6	40.27	48058.86	0.59	0.08	1.07
7	53.30	51154.18	0.58	0.12	1.11
8	62.79	54031.58	0.57	0.20	1.14
9	73.58	56710.74	0.58	0.32	1.17
10	100.00	61111.34	0.50	0.52	1.13

evaluate. The values of other parameters are chosen randomly for the 100 sets as mentioned in Sect. 5.8.

Once the benchmarks are generated, the post-mapping area is compared for the abovementioned synthesis flows. Table 5.2 shows the comparison of the post-mapping area using synthetic benchmarks. The mean value is calculated over the entire benchmark set in the following way. For a particular value of the self-duality index, the mean is calculated over the 100 benchmarks generated. For example, the second column shows the mean of self-duality density calculated for each of the 100 benchmark sets corresponding to a given particular value of the self-duality index. The first two columns show the benchmark's self-duality index (*index*) and the corresponding self-duality density after the logic optimization step of `rewrite;resub`. As mentioned in Algorithm 5.2, *sd-index* corresponds to self-duality density. Hence, for *sd-index* = 10, the SD-density is 100%.

The next four columns in Table 5.2 show the *mean* of the post-mapping area over 100 benchmarks for a particular value of self-duality index for four different flows. The column *baseline* shows the mean post-mapping area for ABC native flow. The next three columns show the area comparison with respect to the baseline results in percentage. It can be noticed that as compared to the baseline, the *mtl_AIG* columns achieve better area results. This can be ascertained due to better mapping results with the technology mapper within the mockturtle framework [Tem+21].

For the XMG-based flow using the ABC technology mapper, better results are obtained only with higher self-duality ratios. This is consistent with the assumption that with higher self-duality in circuits, the XMG-based flow achieves better area results. The last column shows area results with the proposed approach. From the table, this flow achieves the best area results for circuits with higher self-duality. The improvement with XMG-based flow, though consistent, is close to $\sim$1% only and is not so significant. This can be ascertained due to the fact that synthetic benchmarks generated using simple graph-based techniques are devoid of irregularities which are present in an actual benchmark [SVC00]. Further, the AIG-based flows are well-established flows, and hence, they can also achieve relatively good optimization. Synthetic benchmarks cannot be taken as the gold standard for any logic synthesis applications, and hence, the next set of experiments are carried over a real benchmark suite. Due to these limitations, we focus our evaluation on real cryptography benchmarks. It is to be noted that synthetic benchmarks are used just to show the impact of XMG-based optimization for RFET-based circuits as other traditional benchmark suites are mostly based on NAND-centric (CMOS) logic.

5.9.4 Cryptographic Benchmark Suite

While the previous experiment was conducted on synthetic benchmarks, an evaluation on cryptographic benchmarks is carried out in this section. These benchmarks are taken from high-level cryptography protocols such as *fully homomorphic encryption* (FHE) and secure *multiparty communication* (MPC) [Alb+15, Cha+17].[3] This benchmark suite contains circuits ranging from block ciphers (AES and DES) and hash functions such as (MDA-5 and SHA) to arithmetic functions (adders and comparators). It is to be noted that the EPFL benchmark suite [AGD15] is not considered in our evaluation as the benchmarks are not representative of the use case. Almost all the benchmarks have poor self-duality density, except for a few benchmarks such as *adder* and *square*.

Runtime Improvement with Filtering in XMG Resubstitution

In order to measure the improvement in runtime using the XOR3-based filtering rule, one iteration of resubstitution (with and without filtering) over cryptographic benchmarks is performed using XMGs. The results are shown in Table 5.3. The third and the fourth columns in Table 5.3 show the runtime for the proposed resubstitution algorithm in both flows. Figure 5.4 shows the percentage improvement in runtime for individual benchmarks. The horizontal line shows the average runtime improvement across all benchmarks. One can see that an average runtime improvement of 59.48% across all benchmarks can be achieved using the proposed filtering rule.

Improving Self-Duality Density

The impact of the proposed algorithms on the self-duality of the circuit is evaluated with this experiment. A simple approach is followed here. As done in all the experiments, the benchmarks are read in AIG and then converted to XMG using

[3] https://homes.esat.kuleuven.be/nsmart/MPC/.

Table 5.3 Runtime for XMG resubstitution with and without the proposed filtering rule

Benchmarks	Size before	Runtime without filter (seconds)	Runtime with filter (seconds)
AES-expanded	25,435	20.19	5.30
AES-non-expanded	31,642	26.16	6.79
DES-expanded	20,199	447.56	37.85
DES-non-expanded	20,177	443.11	35.03
adder_32bit	99	0.01	0.01
adder_64bit	203	0.03	0.02
comparator_32_s_lt	119	0.03	0.02
comparator_32_s_lt	129	0.05	0.02
comparator_32_uns_lt	119	0.04	0.01
comparator_32_uns_lt	129	0.06	0.02
md5	27,867	11.31	4.34
mult_32x32	5378	4.30	0.84
sha-1	39,426	8.96	4.78
sha-256	83,381	22.84	12.48

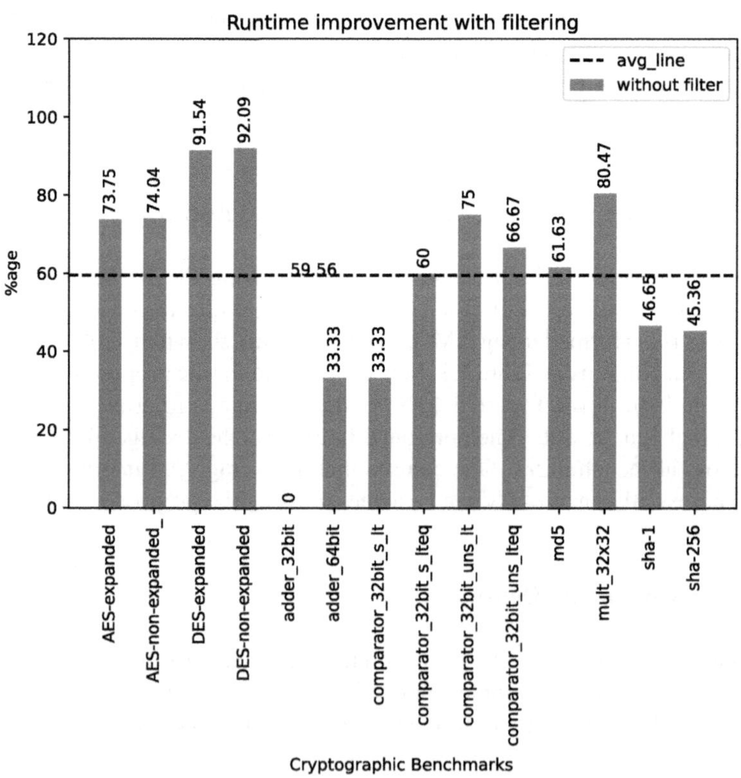

Fig. 5.4 Runtime improvement in XMG resubstitution using the proposed filtering rule

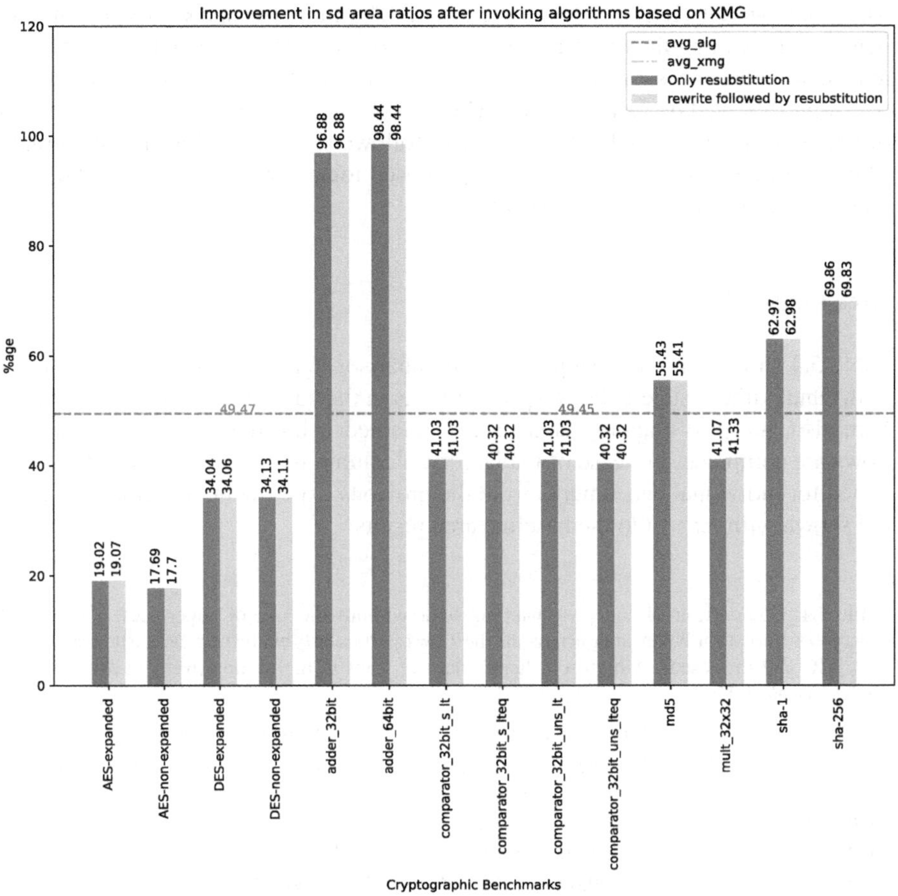

Fig. 5.5 Improvement of self-duality ratio after calling only resub, rewrite followed by resub. All the algorithms are called until convergence. Self-duality almost remains the same for both these function calls—only resubstitution (49.45%) and rewriting followed by resubstitution (49.47%). That is why the two lines almost overlap

the *node resynthesis* technique. Then, algorithm(s) are invoked iteratively until no size improvement of more than 5% is possible. For each of the benchmarks, the self-duality density for the XMG logic graph is calculated. The results are shown in Fig. 5.5. The bars show the self-duality density for the logic graphs using different algorithms. The first bar *initial sd ratio* shows the initial self-duality density calculated on the obtained XMG logic graph of individual benchmarks before invoking the algorithms. The self-duality density does not change when invoking only the rewriting algorithm because the NPN-based rewriting retains the number of MAJ and 2-input XOR nodes. However, as visible in the second bar, resubstitution leads to a significant increase in the self-duality of individual circuits. This can be ascertained due to the addition of extra self-dual nodes (3-input

MAJ and 3-input XORs) to the logic graph. The third bar shows the self-duality density after calling the two algorithms sequentially. It can be seen that self-duality almost remains the same for both these function calls—only resubstitution and rewriting followed by resubstitution. This can be observed in the two average lines (49.45% for resub and 49.47% for rewrite followed by resub) drawn in the graph. This experiment demonstrates that the proposed logic synthesis algorithm helps to increase the self-duality of the circuit.

Area Comparison

Table 5.4 shows the post-mapping area comparison for cryptographic benchmarks with[4] and without supergates, respectively. As in the case of synthetic benchmarks, here also, the post-mapping area results obtained using the aforementioned four flows are compared. For ease of readability, a column for sd density value has been added for individual benchmarks from the previous experiment. The improvements are shown with respect to the baseline area results.

Table 5.4 Comparison of the post-mapping area without the use of supergates. The use of supergates leads to a better area across all the flows particularly with the mockturtle framework (mtl_AIG and Proposed). A high self-duality density leads to higher improvement over baseline for XMG-based flows

		Without supergates				With supergates			
Benchmarks	SD density	Baseline rwrs area	xmg _rwrs %	mtl _aig %	Prop. (%)	Baseline rwrs area	xmg _rwrs %	mtl _aig %	Prop. (%)
AES-exp.	19.07	72231.00	−3.50	−4.98	−4.03	71568.00	−2.77	−1.34	−2.97
AES-non-exp.	17.7	92190.00	−1.29	−4.11	−1.26	90962.50	−1.32	−1.08	−1.26
DES-exp.	34.06	35419.00	−1.52	−8.44	−15.44	35322.00	−1.32	−2.03	−7.14
DES-non-exp.	34.11	35495.50	−0.94	−7.88	−15.75	35230.00	−1.27	−2.00	−7.72
adder_32bit	96.88	270.00	−10.74	0.00	0.00	270.00	0.00	0.00	0.00
adder_64bit	98.44	542.00	−11.25	0.00	0.00	542.00	0.00	0.00	0.00
comp._s_lt	41.03	246.00	2.44	−22.97	−7.93	235.00	6.81	0.64	**6.17**
comp._s_lteq	40.32	235.00	−4.47	−22.55	−27.23	225.50	−1.11	−8.87	−5.10
comp._uns_lt	41.03	246.00	0.81	−22.97	−7.72	235.00	5.11	0.64	**6.60**
comp._uns_lteq	40.32	235.00	−4.47	−22.55	−27.23	225.50	−1.11	−8.87	−5.10
md5	55.41	78489.00	2.20	5.81	**8.24**	79810.00	2.07	7.58	**9.93**
mult_32x32	41.33	9917.50	2.39	−7.34	−15.95	10208.00	0.74	−2.65	**3.92**
Sha-1	62.98	116425.00	1.17	7.72	**9.19**	118003.50	0.79	9.12	**10.61**
Sha-256	69.83	232028.50	−0.56	10.28	**9.55**	236238.00	0.84	12.08	**12.36**

[4] Using ABC command to create supergate library: `super -I 5 -L 3 -N 0 -T 1 -D 0.00 -A 0.00`. The command in ABC generates 5-input supergates by combining three levels of primitive gates. This command generates a total of ~17000 supergates for mapping.

One can notice that most of the benchmarks from the cryptography domain have a high density of self-dual gates (>50%), particularly parity functions as it is an integral logic function in any cryptographic applications. In case of benchmarks, where the sd ratio is lower (<50%) (*DES*, *comparator*), the AIG-based approach gives better results. Hence, both AIG-based approaches (*baseline* and *mtl_AIG*) achieve better area results for such benchmarks. This is due to the fact that AIG representation consists of 2-input nodes, and hence, more numbers of smaller cuts are available for mapping compared to XMGs which consist of bigger XOR and MAJ nodes. Hence, the mapping achieves better optimization in terms of area-oriented mapping for these circuits. However, for benchmarks, *md5*, *SHA-1*, and *SHA-256*, the proposed XMG-based approach (*proposed*) outshines other synthesis flows and gives superior results (shown in bold in Table 5.4). The higher self-duality in these circuits leads to better preservation of self-dual logic using XMGs. This results in the mapping of more self-dual logic gates leading to improved area results for RFET-based circuits as explained in Sect. 5.5.

Consideration of supergates leads to even better results as it can mitigate structural bias issues [Cha07]. However, in case of ABC-based flows (*baseline* and *abc_xmg*), the mapper leads to poor area results for some benchmarks. This can be ascertained as the mapper can get stuck in a local minimum as area-oriented mapping is an intractable problem and is driven by heuristics. However, mockturtle-based flows (mtl_AIG and proposed) are more consistent here as using supergates leads to more uniform area reduction (*comparator*, *multiplier*, *SHA-1*, *SHA-256*). Using supergates with the proposed approach achieves best area results of up to 12.36% for circuits with higher self-duality. This gives an experimental evidence that for circuits with higher self-duality, the XMG-based approach obtains better area results for RFET-based circuits.

Exploring Why Higher Self-Duality Density Leads to Better Area Results with XMG-Based Approach

In this set of experiments, exploration is done to reason why XMG-based approaches lead to better area reduction as compared to AIG-based approaches for circuits with higher self-duality density. The mockturtle-based synthesis flows are used for this experiment (mtl_AIG and proposed). Three important metrics are calculated—the total number of cuts available during mapping, the number of total self-dual cuts, and the number of nontrivial self-dual cuts. These metrics give an overall idea regarding what kind of cuts are available during mapping. Additionally, from post-mapping results, the ratio of the area of self-dual logic gates to the total area of the circuit is computed. Table 5.5 shows these values for both AIG- and XMG-based flows. For each benchmark, the following metrics are shown—the total cuts, the ratio of self-dual cuts to the total cuts, the ratio of nontrivial self-dual cuts to the total cuts, and finally, the ratio of area contribution from self-dual logic gates to the total post-mapping area.

Table 5.5 Comparison of ratios of self-dual cuts between AIG- and XMG-based representation

	AIG			XMG		
Benchmark	Total cuts	sd cuts (%)	sd3 cuts (%)	Total cuts	sd cuts (%)	sd3 cuts (%)
AES-expanded	2906133	5.85	4.74	363829	24.50	18.92
AES-non-expanded	3754949	6.20	4.94	561621	29.12	23.40
DES-expanded	1109707	0.82	0.48	292045	8.07	4.92
DES-non-expanded	1117971	0.83	0.48	291256	8.08	4.92
adder_32bit	4591	3.81	1.31	127	96.06	47.24
adder_64bit	9576	3.76	1.29	255	98.04	48.63
comparator_32_s_lt	4084	1.15	0.44	564	14.54	4.96
comparator_32_s_lteq	3668	1.04	0.35	622	13.99	5.31
comparator_32_uns_lt	4084	1.15	0.44	556	14.93	5.04
comparator_32_uns_lteq	3668	1.04	0.35	622	13.99	5.31
md5	3395540	5.24	3.97	168548	37.75	27.01
mult_32x32	285268	3.01	2.12	39800	12.51	6.97
sha-1	5632714	6.62	5.38	174914	47.60	35.18
sha-256	11774568	6.22	5.02	370611	51.87	38.39
Average	2002056.33	3.31	2.16	151104.07	35.80	20.72

From Table 5.5, the number of total cuts in AIG is more than that in XMG as AIG uses a 2-input AND node compared to XMG that uses larger 2-input XOR and 3-input MAJ nodes. However, the ratio of self-dual cuts is much higher in the case of XMG as compared to AIG. An interesting observation is that this ratio is also much higher in case of XMGs for circuits that have a higher self-duality density (*sha-1*, *sha-256*, *md5*). It clearly indicates that XMGs offer more self-dual cuts during mapping compared to AIGs. Similarly, for circuits with higher self-duality density, XMG on average has a higher *sd area ratio* (54.38% vs 46.72%) compared to AIGs as shown in Fig. 5.6. It can be noticed that the XMG-based approach has higher *sd area ratio* for all the benchmarks. Due to the higher *sd area ratio*, XMGs lead to better area reduction for RFET-based implementation compared to AIGs for circuits with higher self-duality.

Another practical benefit with the XMG-based flow, which can be correlated with the results shown in Table 5.5 and was earlier demonstrated in [Haa+17], is the improvement in the runtime for XMG-based flow. XMG-based flow has to iterate over fewer cuts as compared to the AIG-based flow which leads to reduction in the overall runtime of the synthesis flows. In terms of runtime to carry out technology-independent mapping for all the cryptography benchmarks, XMG takes in total 7.06 s as compared to AIG that takes 59.88 s.

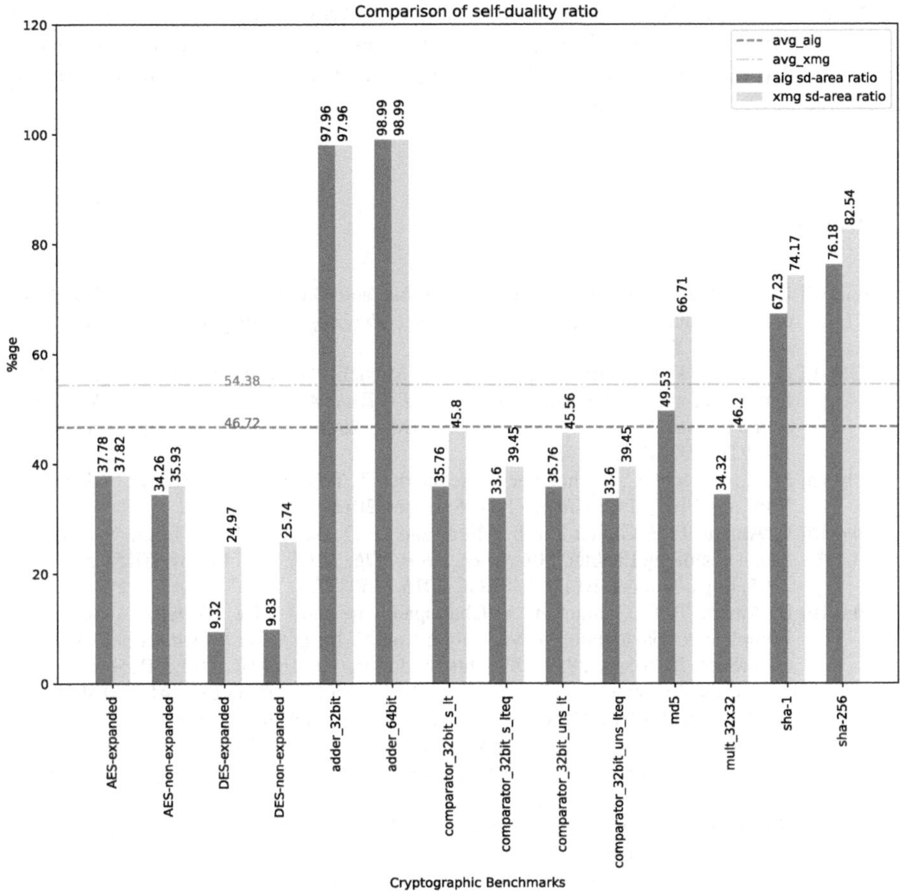

Fig. 5.6 Comparison of final self-duality area ratio between AIG and XMG

5.10 Concluding Remarks and Future Research Directions

The present chapter explores both logic synthesis and technology mapping from an emerging technology perspective. With the particular aim of preserving self-duality in circuits, an XMG-based logic synthesis solution for RFET-based circuits is investigated. XMGs are used for two reasons: (i) XMGs are a compact representation for both unate and binate logic; (ii) the logic primitives in XMGs (MAJs and 2-input XORs) can efficiently represent self-dual logic gates because both Majority and odd-input XORs are self-dual. Additionally, with the logic representation-agnostic versatile mapper, the limitation of previous work [Rai+21] of converting XMGs into AIGs before technology mapping has been resolved. In comprehensive experimental evaluations, the XMG-based flow is compared with three different sets of experiments. It has been shown that for circuits with higher self-duality, the XMG-based synthesis achieves better area results. Future work directions include the development of more optimization algorithms for XMGs. Additionally, measures to deal with the problem of structural bias within technology mapping have to be also explored.

References

[AGD15] Luca Amarú, Pierre-Emmanuel Gaillardon, and Giovanni De Micheli. "The EPFL combinational benchmark suite". In: *Proceedings of the 24th International Workshop on Logic & Synthesis (IWLS)*. 2015.

[AGM16] L. Amarú, P. E. Gaillardon, and G. De Micheli. "Majority-Inverter Graph: A New Paradigm for Logic Optimization". In: *IEEE Transactions on Computer-Aided Design of Integrated Circuits and Systems* 35.5 (May 2016), pp. 806–819. https://doi.org/10.1109/TCAD.2015.2488484.

[Alb+15] M. R. Albrecht, C. Rechberger, T. Schneider, T. Tiessen, and M. Zohner. "Ciphers for MPC and FHE". In: *EUROCRYPT*. Springer. 2015.

[Ama+15] L. Amarú, P. E. Gaillardon, S. Mitra, and G. De Micheli. "New Logic Synthesis as Nanotechnology Enabler". In: *Proceedings of the IEEE* 103.11 (Nov. 2015), pp. 2168–2195. https://doi.org/10.1109/JPROC.2015.2460377.

[Ama+16] L. Amarú, P. E. Gaillardon, A. Chattopadhyay, and G. De Micheli. "A Sound and Complete Axiomatization of Majority-n Logic". In: *IEEE Transactions on Computers* 65.9 (Sept. 2016), pp. 2889–2895. https://doi.org/10.1109/TC.2015.2506566.

[Ama+18] Luca Gaetano Amarù, Mathias Soeken, Patrick Vuillod, Jiong Luo, Alan Mishchenko, Janet Olson, Robert K. Brayton, and Giovanni De Micheli. "Improvements to Boolean resynthesis". In: *DATE*. 2018. https://doi.org/10.23919/DATE.2018.8342108.

[Ama17] Luca Gaetano Amaru. *New Data Structures and Algorithms for Logic Synthesis and Verification*. Springer, 2017.

[BM10] Robert Brayton and Alan Mishchenko. "ABC: An academic industrialstrength verification tool". In: *Computer Aided Verification*. Springer. 2010.

[Cha+05] S. Chatterjee, A. Mishchenko, R. Brayton, X. Wang, and T. Kam. "Reducing structural bias in technology mapping". In: *ICCAD-2005. IEEE/ACM International Conference on Computer-Aided Design, 2005*. Nov. 2005, pp. 519–526. https://doi.org/10.1109/ICCAD.2005.1560122.

[Cha+17] Melissa Chase, David Derler, Steven Goldfeder, Claudio Orlandi, Sebastian Ramacher, Christian Rechberger, Daniel Slamanig, and Greg Zaverucha. "Post-quantum zero-knowledge and signatures from symmetric-key primitives". In: *ASCCCS*. 2017.

[Cha07] S. Chatterjee. *On algorithms for technology mapping (Doctoral Thesis)*. University of California, Berkeley, 2007.

[Chu+19] Zhufei Chu, Mathias Soeken, Yinshui Xia, Lunyao Wang, and Giovanni De Micheli. "Structural Rewriting in XOR-Majority Graphs". In: *Proceedings of the 24th Asia and South Pacific Design Automation Conference*. ASPDAC '19. Tokyo, Japan: Association for Computing Machinery, 2019, pp. 663–668. https://doi.org/10.1145/3287624.3287671.

[CMB08] Kai-Hui Chang, Igor L. Markov, and Valeria Bertacco. "Fixing Design Errors With Counterexamples and Resynthesis". In: *IEEE TCAD* (2008). https://doi.org/10.1109/TCAD.2007.907257.

[CWD99a] Jason Cong, Chang Wu, and Yuzheng Ding. "Cut Ranking and Pruning: Enabling a General and Efficient FPGA Mapping Solution". In: *ISFPGA*. 1999. https://doi.org/10.1145/296399.296425.

[CWD99b] Jason Cong, Chang Wu, and Yuzheng Ding. "Cut ranking and pruning: Enabling a general and efficient FPGA mapping solution". In: *Proceedings of the 1999 ACM/SIGDA seventh international symposium on Field programmable gate arrays*. 1999, pp. 29–35.

[Haa+17] W. Haaswijk, M. Soeken, L. Amarù, P. E. Gaillardon, and G. De Micheli. "A novel basis for logic rewriting". In: *2017 22nd Asia and South Pacific Design Automation Conference (ASP-DAC)*. Jan. 2017, pp. 151–156. https://doi.org/10.1109/ASPDAC.2017.7858312.

[HFS17] I. H'alecek, P. Fišer, and J. Schmidt. "Are XORs in logic synthesis really necessary?" In: *2017 IEEE 20th International Symposium on Design and Diagnostics of Electronic Circuits Systems (DDECS)*. Apr. 2017, pp. 134–139. https://doi.org/10.1109/DDECS.2017.7934583.

[Hua+13] Zheng Huang, LingliWang, Yakov Nasikovskiy, and Alan Mishchenko. "Fast Boolean matching based on NPN classification". In: *FPT*. 2013. https://doi.org/10.1109/FPT.2013.6718374.

[KGP01] Andreas Kuehlmann, Malay K. Ganai, and Viresh Paruthi. "Circuit- Based Boolean Reasoning". In: *Proceedings of the 38th Annual Design Automation Conference*. New York, NY, USA, 2001, pp. 232–237. https://doi.org/10.1145/378239.378470.

[KK04] Victor N. Kravets and Prabhakar Kudva. "Implicit enumeration of structural changes in circuit optimization". In: *DAC*. 2004. https://doi.org/10.1145/996566.996691.

[MB06] A. Mishchenko and R. K. Brayton. "Scalable logic synthesis using a simple circuit structure". In: *IWLS*. 2006.

[MBV06] Valavan Manohararajah, Stephen Dean Brown, and Zvonko G Vranesic. "Heuristics for area minimization in LUT-based FPGA technology mapping". In: *IEEE Transactions on Computer-Aided Design of Integrated Circuits and Systems* 25.11 (2006), pp. 2331–2340.

[MCB06] A. Mishchenko, S. Chatterjee, and R. K. Brayton. "DAG-aware AIG rewriting a fresh look at combinational logic synthesis". In: *DAC*. 2006. https://doi.org/10.1145/1146909.1147048.

[Mic94] Giovanni De Micheli. *Synthesis and optimization of digital circuits*. 1st. McGraw-Hill, 1994, p. 579.

[Mis+07] Alan Mishchenko, Sungmin Cho, Satrajit Chatterjee, and Robert Brayton. "Combinational and sequential mapping with priority cuts". In: *2007 IEEE/ACM International Conference on Computer-Aided Design*. IEEE. 2007, pp. 354–361.

[Mis+11] Alan Mishchenko, Robert K. Brayton, Jie-Hong R. Jiang, and Stephen Jang. "Scalable don't-care-based logic optimization and resynthesis". In: *ACM TRECTS*. (2011). https://doi.org/10.1145/2068716.2068720.

[MM90] F Mailhot and G De Micheli. "Technology Mapping Using Boolean Matching and Don'T Care Sets". In: *Proceedings of the Conference on European Design Automation*. EURO-DAC '90. Glasgow, Scotland: IEEE Computer Society Press, 1990, pp. 212–216.

[Net+19] Walter Lau Neto, Vinicius N Possani, Felipe S Marranghello, Jody Maick Matos, Pierre-Emmanuel Gaillardon, Andre Inacio Reis, and Renato Perez Ribas. "Exact benchmark circuits for logic synthesis". In: *IEEE Design & Test* 37.3 (2019), pp. 51–58.

[Rai+19] S. Rai, J. Trommer, M. Raitza, T. Mikolajick, W. M. Weber, and A. Kumar. "Designing Efficient Circuits Based on Runtime-Reconfigurable Field-Effect Transistors". In: *IEEE Transactions on Very Large Scale Integration (VLSI) Systems* 27.3 (Mar. 2019), pp. 560–572. https://doi.org/10.1109/TVLSI.2018.2884646.

[Rai+20] S. Rai, M. Raitza, S.S. Sahoo, and A. Kumar. "DiSCERN: Distilling Standard-Cells for Emerging Reconfigurable Nanotechnologies". In: *DATE*. 2020.

[Rai+21] Shubham Rai, Heinz Riener, Giovanni De Micheli, and Akash Kumar. "Preserving Self-Duality During Logic Synthesis for Emerging Reconfigurable Nanotechnologies". In: *2021 Design, Automation Test in Europe Conference Exhibition (DATE)*. 2021, pp. 354–359. https://doi.org/10.23919/DATE51398.2021.9474112.

[Rie+18] Heinz Riener, Eleonora Testa, Luca G. Amarù, Mathias Soeken, and Giovanni De Micheli. "Size Optimization of MIGs with an Application to QCA and STMG Technologies". In: *NANOARCH*. 2018. https://doi.org/10.1145/3232195.3232202.

[Rie+19a] Heinz Riener, Winston Haaswijk, Alan Mishchenko, Giovanni De Micheli, and Mathias Soeken. "On-the-fly and DAG-aware: Rewriting Boolean Networks with Exact Synthesis". In: *DATE*. 2019. https://doi.org/10.23919/DATE.2019.8715185.

[Rie+19b] Heinz Riener, Eleonora Testa, Winston Haaswijk, Alan Mishchenko, Luca G. Amarù, Giovanni De Micheli, and Mathias Soeken. "Scalable Generic Logic Synthesis: One Approach to Rule Them All". In: *DAC*. 2019. https://doi.org/10.1145/3316781.3317905.

[RMS20] Heinz Riener, Alan Mishchenko, and Mathias Soeken. "Exact DAGaware Rewriting". In: *DATE*. 2020.

[Soe+18] Mathias Soeken, Heinz Riener, Winston Haaswijk, Eleonora Testa, Bruno Schmitt, Giulia Meuli, Fereshte Mozafari, and Giovanni De Micheli. "The EPFL logic synthesis libraries". In: *arXiv preprint arXiv:1805.05121* (2018).

[SVC00] D. Stroobandt, P. Verplaetse, and J. van Campenhout. "Generating synthetic benchmark circuits for evaluating CAD tools". In: *IEEE Transactions on Computer-Aided Design of Integrated Circuits and Systems* 19.9 (2000), pp. 1011–1022. https://doi.org/10.1109/43.863641.

[Tem+21] Alessandro Tempia Calvino, Shubham Rai, Heinz Riener, Giovanni De Micheli, and Akash Kumar. "From logic to gates: A versatile mapping approach to restructure logic". In: *Proceedings of the 30th International Workshop on Logic & Synthesis (IWLS 2021)*. 2021.

Chapter 6
Physical Synthesis Flow and Liberty Generation

Physical synthesis is a crucial step in Electronic Design Automation (EDA) to derive actual physical circuits from a technology-mapped netlist of a Boolean logic network. It brings out important consideration of actual physical geometry and the area of standard cells while finalizing a given circuit. Other factors such as wirelengths, parasitics, capacitances, and other electrical effects are also computed to give close estimates of the various physical parameters such as area, delay, and power for a given circuit. Particularly, for emerging technologies, it is an important step to analyze and compare the actual physical benefits of the technology in comparison to an existing and established Complementary Metal Oxide Semiconductor (CMOS) technology. While an emerging technology can be promising at a transistor level, a benchmark-level evaluation in terms of post-physical synthesis (area, power, and delay numbers) is necessary to understand the feasibility aspects of such an emerging technology. It also provides feedback to the technology model engineer to bring forth the limitations of the current models and explore ways to improve existing technology models for emerging technologies.

This chapter proposes a physical synthesis flow for circuits based on reconfigurable technologies. Using a target Silicon Nanowire (SiNW)-based Reconfigurable Field-Effect Transistor (RFET) technology, the chapter explores and evaluates RFET-based circuits in terms of multiple physical parameters. Further, the chapter investigates innovative physical synthesis approaches that can benefit RFETs-based circuits in terms of security evaluation. The chapter specifically targets the following research question:

Research Question Considering RFETs to be still in a nascent stage in terms of their maturity compared to CMOS, how feasible is it to make actual physical circuits based on RFETs?

With the above research question in mind, the chapter discusses the major challenges and opportunities that need to be explored for enabling efficient physical circuits based on RFETs. This chapter presents a tailor-made physical synthesis flow by carrying out a benchmark-level evaluation of RFETs-based circuits.

S. Rai, A. Kumar, *Design Automation and Applications for Emerging Reconfigurable Nanotechnologies*, https://doi.org/10.1007/978-3-031-37924-6_6

6.1 Contributions

The major contributions of this chapter are as follows:

1. Layouts for a set of logic gates based on SiNW have been proposed. A reconfigurable ready layout for the *Minority* logic gate is also demonstrated.
2. A physical synthesis flow using both open-source tools and the industrial *Cadence Encounter* has been proposed. Particularly, `.lef` and `.lib` files are made available using the 22-nm technology model for SiNW RFETs. Both the flows are available under open-source license at (https://github.com/shubhamrai26/repopro).
3. Using logic locking as a target security application, two approaches for placement and routing have been proposed for RFETs-based circuits. The concept of driver cells is introduced to generate the P and $\overline{P}$ signals required by the reconfigurable logic gates. Such an approach leads to localization of reconfigurable components of the circuit that can reduce area overheads for logic locking schemes in RFET-based circuits [Rai+20b, Sin+21, Kne20].
4. Since RFETs are used for reconfigurable logic gates and reconfigurable circuits [Rai+19b], an investigation into how the ratio between reconfigurable and conventional logic gates affects chip area, wirelength consumption, and power consumption has been carried out.

In order to create a Verilog-A model and to extend it to generate `.lef` and `.lib` files, a table model is created that uses the 22-nm technology model. The technology model is proposed to pattern the individual nanowire width and to define the half-pitch between parallel arranged nanowire channels [Bal+17]. The target technology is the fully symmetrical RFET as proposed in [Hei+12] and adapted to an Silicon-On-Insulator (SOI) platform. Both open-source and industrial flows are demonstrated to show the comparison with an equivalent CMOS model.

6.2 Organization

The present chapter is organized as follows: Sect. 6.3 discusses the related work in the domain of physical synthesis for RFETs and the motivation behind exploring different physical synthesis approaches for RFETs. Then, Sect. 6.4 introduces the SOI platform for SiNW-based RFETs. This is followed by Sect. 6.5 which introduces the layouts for both static and reconfigurable logic gates based on RFETs. Section 6.6 introduces the table model and the complete tool flow used to generate the `.lef` and `.lib` files for SiNW RFET technology. The concept of driver cells and new techniques for physical synthesis is explained in Sect. 6.7. Experiments are described in Sect. 6.8 which is followed by discussions in Sect. 6.9. Concluding remarks are given in Sect. 6.10.

6.3 Background and Related Work

6.3.1 Related Works

As discussed in Chap. 2, RFETs, by the virtue of their device-level reconfiguration, provide an alternative path to increase the number of functions offered per unit transistor. In spite of its great promise, most of the work in enabling EDA for RFETs has been mainly at the logic level as discussed in [Tro+16, Rai+17, Gai+14] and [Gai+13]. While [Tro+16] demonstrated efficient and exemplary logic gates, [Rai+17] and [Gai+14] showed the potential of these RFETs in bigger circuits and Silicon-On-Chip (SOC) core components, respectively. In terms of the synthesis flow, most of the earlier works targeted either at logic optimization [AGM16, Rai+18, Rai+20a] or focused on manually designed circuit elements [ZGD13, Gai+14, AGM13]. A major research aspect that has been missing until now is the support of these works from the physical synthesis point of view, i.e., their compatibility with existing synthesis flows.

There have been relatively fewer attempts to target physical synthesis flows for RFETs-based circuits [BM15, RRK18, Reu+21]. Earlier works like [Bob+12, BM15] focused on physical synthesis for RFETs on a configurable sea-of-tiles-like fabric. The work done in [Bob+12, BM15] addresses the routing congestion arising from the extra gate terminal in the SiNWs RFET [Bob+12]. However, their approach was technology independent and they showed layouts for basic logic gates only. Additionally, such a sea-of-tiles fabric is incompatible with conventional physical synthesis flows. For RFETs, where a single device has more contact terminals as compared to the traditional CMOS, placement and routing are more challenging for conventional approaches. Although RFETs lose to CMOS in terms of the number of contacts per transistor, they make up for this with their higher functional expression, since the overall number of transistors per circuit is greatly reduced. This has been demonstrated particularly in [Rai+19b, Rai+17].

None of these works have tried to demystify the entire design flow—logic synthesis up to physical synthesis for SiNW RFETs. Neither is there an evaluation of SiNW through a parallel CMOS standard flow from a technology perspective. This chapter is aimed at formalizing a complete physical synthesis flow along with layouts for various logic gates based on SiNW RFETs. The chapter also focuses on optimizing the post-placement area by exploring ways to improve the physical synthesis for RFET-based circuits through the creative use of multiple *power domains* and *power shutoff* (PSO). There has been no prior work on the placement and routing of reconfigurable logic gates based on these emerging technologies, which is a prerequisite for reconfigurable logic circuits. Hence, it is required to design a complete EDA flow comprising both logic and physical synthesis for RFET-based circuits.

6.3.2 Motivation

Due to the inherent ambipolarity offered by these emerging reconfigurable nanotechnologies, transistors and hence logic gates demonstrate runtime reconfigurability by applying different bias voltages to the program gate of a single or a group of transistors [Rai+19b]. This runtime reconfigurability can be exploited for developing polymorphic logic gates [Mac11] at low area, power, and delay overheads. Such polymorphism allows achieving efficient solutions for hardware security, particularly for logic locking schemes [Bi+16, Rai+19a, Rai+20b].

However, earlier works from a hardware security perspective have overlooked the physical design flow for circuits based on RFETs. Hence, the impact of security measures over routing and placement has been missing which leads to more abstract overheads in terms of area and power. This chapter looks at physical synthesis to support such logic-locking-based hardware security schemes. Various approaches have been explored in physical synthesis flow for RFETs-based circuits and discuss concrete area and power numbers by carrying out a benchmark-level evaluation. A discussion regarding the variability of reconfigurable logic and its effects on the area and other parameters of the circuit has been done as this is relevant from a security point of view. The present work deals with evaluations and analysis in terms of exploring efficient physical synthesis flows for reconfigurable FET-based circuits. The security analysis is avoided in this chapter as these security evaluations for logic-level modeling have been carried over RFETs-based circuits in the next chapter. The next chapter demonstrates that the practical security can easily be achieved against seminal SAT-based attacks [SRM15] using RFET-based circuits. This chapter, thus, focuses on the backend flow of circuit design in terms of physical synthesis. The security evaluations using logic-level modeling have been carried over RFETs-based circuits in Chap. 7.

6.4 Silicon Nanowire Reconfigurable Transistors

For this work,[1] an SOI-based 22-nm technology is proposed to pattern the minimal width of individual nanowire ribbons and to define the half-pitch between parallel arranged nanowires. These features can be achieved with a conventional 45-nm technology node and a spacer pattern transfer technology as commonly done for *fin* patterning in FinFETs. The top Si thickness is set to 6.0 nm and the Effective Gate Oxide Thickness (EOT) is set to 1.0 nm. To accommodate the two parallel gates per device, the technologically simplest route is chosen with a gate pitch of $\sim$110 nm and an active channel length of $\sim$100 nm. This would imply a gate composition and patterning flow as similar as possible to the commonly employed MOSFET gates

[1] The model was developed in collaboration with NaMLaB.

in CMOS. As only a single kind of transistor is needed and no doping is required, at least one lithography step and four implantation steps with associated dopant activation anneals are spared compared to a 45-nm CMOS technology. Nevertheless, it can be noted that the transistor cells for the RFETs can be defined in a much more compact way, e.g., by self-alignment as previously shown in [Mar+12]. However, this requires an additional lithographic step compared to the mentioned approach but allows in a channel length shrinkage from 100 nm to ∼48 nm or less. Since RFET devices exhibit Schottky junctions, they introduce an energy barrier even in the on-state; therefore, drive currents are lower than in conventional MOSFETs for comparable dimensions. Accordingly, they are slower as compared to CMOS-based devices but they have substantially lower static leakage power consumption.

Germanium nanowires as implemented in RFETs [Tro+17a] offer a similar structure but better electrical properties. Therefore, they are able to outperform previous polarity-controllable device concepts on other material systems in terms of minimized threshold voltages and higher normalized on-currents. The present work lays the foundation for emerging reconfigurable technologies with different channel materials.

6.5 Layouts for Logic Gates

Figure 6.1 shows the proposed layouts for logic gates—*INV*, *2-NAND*, and *3-MIN* based on SiNW dual-gate RFETs exclusively. For gate-level schematics of these layouts, readers are advised to go through Chap. 3. It is to be noted that only dual-gate RFETs (DIGFET) have been considered for developing these layouts. While Verilog-A models for RFETs with three-input all-around gate terminals (TIGFETs) have been developed [Gor+19], at the time of writing this thesis, efforts were still undergoing to design the most optimal layouts for these TIGFETs. Pitch considerations between various gate terminals have to be taken care of so as to have the correct pin density for optimal place and route during the physical synthesis

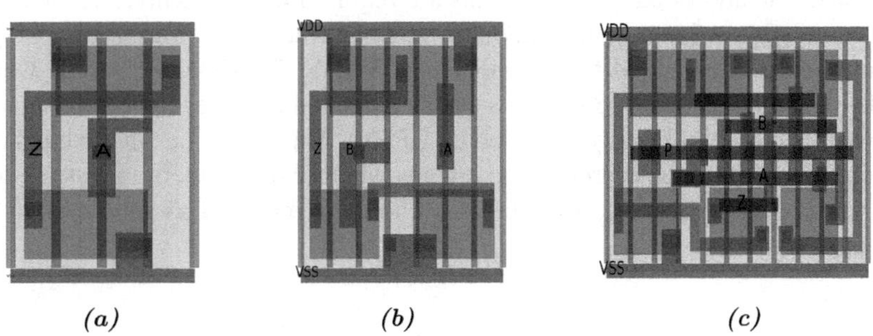

Fig. 6.1 Layouts of logic gates. The first two layouts are for static logic gates, while the third one is for the reconfigurable MIN logic gate. (**a**) Inverter. (**b**) 2-NAND. (**c**) MINORITY

process. RFETs with multi-input gate terminals (MIGFET) though experimentally shown at the transistor level [Zha+14] are still under active research to develop trustworthy models that can produce reproducible results. Multi-input gate FETs like TIGFETs and MIGFETs exhibit a higher potential for minimizing the transistor count and saving the overall area as shown in Chap. 3. Hence, technology models based on TIGFETs or MIGFETs are worth exploring as they can potentially lead to further compaction and better performance results.

6.5.1 Layouts for Static Functional Logic Gates

Figure 6.1a and b shows the layouts of a SiNW RFETs-based NAND and INV logic gates, respectively, predefined in a static manner by the interconnect design. The vertical brown lines in the layout are the gate terminals that are self-aligned to the Schottky contacts below. The three green squares (two in INV) refer to the active regions. A nanowire conducting channel is formed between two vertical gate lines over the silicon well area. With the help of vias, the connections are made to V_{dd}, V_{ss}, $output\ Z$, and $inputs\ A$ and B. This particular layout is static, i.e., the functionality is defined by the layout, and reconfigurability at runtime is not used. One can notice in the NAND layout that the program gates are connected to V_{DD} or V_{SS} internally through vias. This increases the area of the layout but it does not create additional routing overhead.

6.5.2 Layout for Reconfigurable Logic Gate

The MINORITY gate layout, as shown in Fig. 6.1c is a reconfigurable-ready layout. The dark brown horizontal lines are the input/output points. Depending upon the value of S, this layout of the MINORITY gate can function as a NAND gate ($S = 0$) or a NOR gate ($S = 1$). This particular layout possesses more numbers of gate terminals. The additional gate terminals are required to realize inverters within the gate boundary so that it can implement reconfigurable functions. This particular layout integrates extended functionality of a 3-MIN, 2NAND, or 2-NOR within a single logic gate layout. Such layouts can be extended for other logic gates to utilize the runtime reconfigurability provided by RFETs. The choice lies in the hand of the circuit designer to choose among the layouts whether to use them in a static design or a runtime reconfigurable design. This choice is modeled as *replacement ratio* in Sect. 6.9

6.6 Table Model for Silicon Nanowire RFETs

The creation of a table model using experimental data for RFETs is a necessary starting step to create `.lef` and `.lib` files. To create a table model, the internal structure and material properties of the SiNW transistor are modeled in a TCAD simulator. This environment also contains an electrical simulator which can execute transient analog simulations. But the electrical simulator is highly constrained in speed and capacity and, therefore, cannot be used for performance evaluations at the digital circuit level. Based on the electrical characterization of SiNW RFETs, a table model has been proposed using the I–V properties of scaled SiNW considering stressors to adjust symmetry [Bal+17].

To enable a realistic estimation of SiNW logic gate timing and area, *Liberty models* are required. Liberty models are a standard format containing a table model for timing, power, and pin properties of digital standard cells and thereby abstracting from the analog electrical behavior and offering the speed required for large-scale system integration. To create this model, a SPICE simulation is executed for each table entry.

There is no direct link from Technology Computer-Aided Design (TCAD) simulations to transistor models for SPICE simulators. Usually, compact models (based on equations) are used as transistor models. They offer the best simulation speed and robustness, but even though these equations are physics-based, they require major efforts to fit the model to the real transistor behavior over the whole operating range. While this compact modeling is necessary for final technology development, in the early evaluation stages of an emerging technology, a more simple table model might offer a better compromise to link transistor design to electrical simulators. Therefore, a table of current, voltage, and capacitances based on DC simulations inside the TCAD environment is exported as a plain `ASCII` file. For the electrical transistor model, this table is read inside a `Verilog-A` module and thereby can be used by every analog SPICE simulator supporting the required `Verilog-A` constructs. Although this approach cannot model dynamic charge distributions inside the transistor channel, the simulation results of a small circuit is comparable in both TCAD and SPICE. It can be verified that the transient delay error between both simulations is within 20% and is thereby a valid approximation for early technology development. The parasitic capacitances for a typical transistor structure, up to *Metal1*, have been included in this model.

For library characterization, *Synopsys SiliconSmart* is used which integrates all required functionality if a cell netlist and transistor model files are provided. The complete library characterization flow is shown in Fig. 6.2. Based on the cell function, it generates all the timing and power arcs and a set of corresponding HSPICE simulations for each table entry. It then extracts the results and writes out a *Liberty* file (`.lib`) and a matching Verilog behavior model. Based on the SiNW RFET technology status, there has been only one operating condition characterized, which is a typical one at a supply voltage of 1.8 V. While these simulations provide timing and power numbers, the area of the SiNW logic cells needs to be extracted

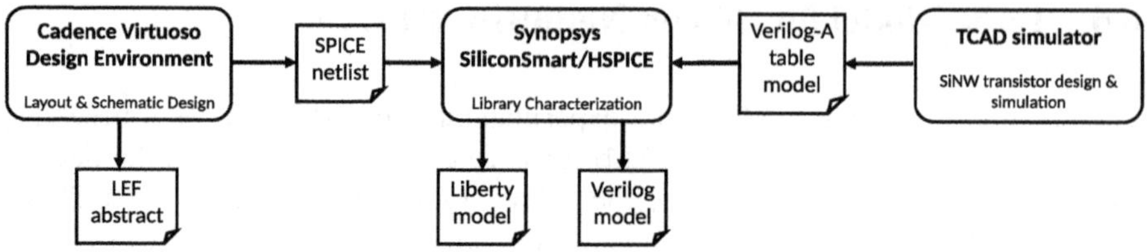

Fig. 6.2 Tool flow for liberty characterization for SiNW-based RFETs

from real cell layouts. As shown in the previous section, these layouts follow advanced node design rules, and the *layout abstract file* (`.lef`) for Place & Route are created by the Cadence Abstract Generator. The `.lef` file contains information like cell boundaries, metal blockages, and pin locations. The area is written to the `.lib` files. The schematics of the cells are exported as SPICE netlists and are used by the characterization tool. Appendices C and D consist of snippets from generated `.lef` and `.lib` files for SiNW-based RFETs, respectively.

6.7 Exploring Approaches for Physical Synthesis

This section covers various physical synthesis approaches for circuits based on RFETs. First, the default flow of standard Placement & Route (P&R) is explained. This flow is also used to compare RFETs-based implementation with the CMOS-based implementation. Second, the concept of *driver cells* is introduced as that is integral for the flows meant for area optimization. Last, the two approaches—*native* and *island* approach—are presented which use driver cells to have a better packing to achieve area reduction. It is to be noted that presently, these two approaches are supported only for the industrial flow as the Power-shut Off (PSO) support is matured in the Cadence Encounter tool flow. These two approaches are targeted towards hardware security ensuring affordable area overheads for RFETs-based circuits. For the sake of the reader's understandability, a trivial example of logic locking is assumed in which the starting RFET-based circuit is only mapped with *NAND* and *NOR* logic gates. Then, a certain percentage of the static logic gates (*NAND* or *NOR*) in a given circuit are replaced with the polymorphic *MINORITY* gate.

6.7.1 Using the Standard Place and Route Flow

Using the library characterization flow as described in the previous section, the `.lef` and `.lib` files are generated for standard cells such as *INV, NAND, NOR, MUX, XOR, and MIN*. These gates contain an inverter (within the gate boundary), so

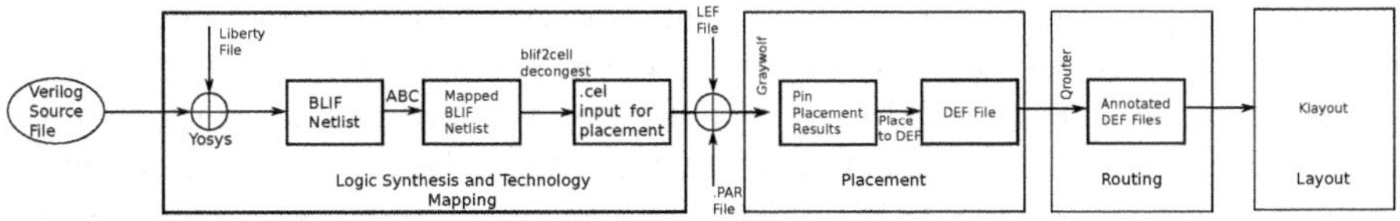

Fig. 6.3 Open-source physical synthesis flow through Qflow [Edw17] Yosys [Wol] ABC [BM10] Graywolf [Gra17] Qrouter [Qro17] Klayout [Köf17]

that both P and $\overline{P}$ are available for reconfiguration. A diagrammatic representation is shown in Fig. 6.6a. The *MIN* gate in Fig. 6.1c is such a layout [RRK18], in which the inverter is located on the left side, connected to V_{dd} and V_{ss}. In this case, M1 metal wires were used for V_{dd} and V_{ss}, while the P and $\overline{P}$ signals were routed together with all other signals using polysilicon, M1, and M2. In this approach, the generated `.lib` and `.lef` are used for carrying out physical synthesis for circuits exclusively based on RFETs. A physical synthesis flow using QFLOW [Edw17] has been shown in Fig. 6.3. Similarly, a flow using industrial tools Cadence Encounter has also been demonstrated. This constitutes the standard P&R flow for circuits based on RFETs involving both static and a reconfigurable gate (MIN).

6.7.2 Open-Source Flow

For the open-source tool flow, *Qflow* [Edw17] is used. The toolchain uses other open-source tools like Yosys [Wol], Graywolf [Gra17], and Qrouter [Qro17] for logic synthesis, placement, and routing steps, respectively. Yosys internally uses the ABC tool [BM10]. The Yosys tool reads the Verilog file and the Liberty file based on the technology as inputs. The ABC tool carries out further logic optimizations and technology-independent mapping. After the mapping stage, the netlist is converted using the *Blif2Cel* tool to generate an input file for the placement tool Graywolf [Gra17]. The tool takes the `.lef` file and the output of the logic synthesis stage to generate a `.cel` file which acts as an input to the Graywolf tool. A `.par` file is generated that uses the technology information from the `.lef`. This file is used by Graywolf. This `.par` file contains the information regarding the routing resources and their constraints. It also contains various controllable parameters for placement. Both the `.par` file and the `.lef` files are fed to the P&R tool. In our flow, the placements of standard cells are taken care of by the Graywolf, and the detailed routing of metal wires is handled by the Qrouter (it is a part of the QFlow toolchain) [Qro17].

Graywolf completes the automatic floor planning and placement and generates an unrouted `.def` file which is then used by Qrouter for further processing. This `.def` file contains all the placement and netlist information. A second iteration for Graywolf is done on the netlist to produce the final placement to weed out minor issues if the placement fails in the first attempt. After the placement of standard cells, the modified netlist is used by the Qrouter which carries out a simple maze routing algorithm for detailed routing. Following this process, we can obtain the area numbers after the placement and routing stage. The routed `.def` file along with the `.lef` file and the .GDSII file for the standard cells can be used to view the final layout. For obtaining the final layout in the .GDSII format, the tool Klayout [Köf17] is used to get the overview of the post-layout circuit.

6.7.3 Concept of Driver Cells

In order to enable reconfigurable functionality using RFET-based logic gates, an inverting logic is imperative to generate $\overline{P}$ from P [Rai+19b]. However, it can be noticed that within a particular circuit, multiple reconfigurable logic gates can share the same value for P. Additionally, since V_{dd} and V_{ss} are only needed for the operation of the internal inverter, their wiring wastes valuable routing resources that cannot be used to connect P, $\overline{P}$, and all other signals. This also leads to an increased pin density and an area overhead for each reconfigurable logic gate in the circuit. These issues can be mitigated by removing the inverter that drives the $\overline{P}$ and P for a reconfigurable logic gate. This leads to a smaller cell area, decreased pin density, and more available routing resources for the *P&R* tool.

Hence, for each group of reconfigurable logic gates sharing the same P and $\overline{P}$, a *driver cell* is instantiated which particularly powers these two signals. These driver cells comprise large inverters with high fan-out, which are usually included in the Process Design Kits (PDKs). This allows us to introduce new layouts of reconfigurable logic gates, as shown in Fig. 6.4b and c, that do not contain an inverter. They need to be connected to P and $\overline{P}$ of a driver cell instead. These layouts further help in reducing the pin density so as to facilitate a better place and route for large circuits. The driver cell is integral to the proposed *native* as well as the *island* approach.

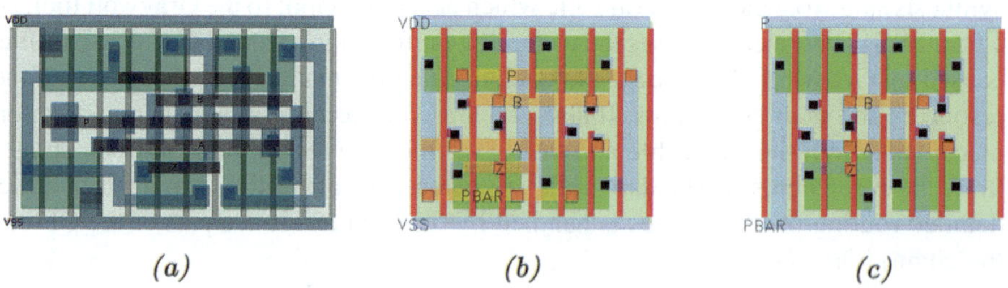

(a) (b) (c)

Fig. 6.4 (**a**) Layout of the minority (*MIN*) logic gate containing an inverter. (**b**) Our newly proposed layout of a *MIN* gate with P, $\overline{P}$ as separate input signals. (**c**) Our improved *MIN* gate layout for the PSO approach

6.7.4 Native Approach

In this approach, reconfigurable logic cells are stripped of their internal inverter as shown in Fig. 6.4b for the *MIN* gate. This leads to a smaller gate area that improves total area and power consumption. The V_{dd} and V_{ss} contacts at the top and bottom are not changed so that the gate can be placed on the default power rails like any other standard cell. However, in these cells, the V_{dd} and V_{ss} contacts are not needed. Instead, the gates are powered by a driver cell using P and $\overline{P}$. This leads to a waste of M1 routing resources, which increases pin density and limits the possible area improvements.

The resulting design flow is shown in Fig. 6.5a. It starts with a Verilog netlist that includes reconfigurable gates. As shown in Fig. 6.6b, a single driver cell powers multiple reconfigurable logic gates using P and $\overline{P}$. Depending on the available drive strengths of the $P/\overline{P}$ drivers, groups with the correct number of reconfigurable gates that share the same $P/\overline{P}$ are created. For each group, a driver is added to the netlist and connected to the individual reconfigurable gates. The driver's input signal P can then be connected to the corresponding pin or signal. Since the new reconfigurable logic gates can be placed and routed like any other standard cell, standard *P&R* as mentioned in Sect. 6.7.1 can be performed to generate the layout.

It can be argued that from a hardware security point of view, sharing of signals can compromise the robustness of a circuit against security attacks. However, in any logic-locked circuit, locking signals are limited by the size of tamper-proof memory [Cha+20], and hence, all of them cannot be possibly made as primary inputs. This promotion of signals from an intermediate signal to primary inputs also adds to routing and area overheads. This particular decision can be left on the hardware designer to explore the amount of sharing; a particular design can afford for maintaining an acceptable level of security.

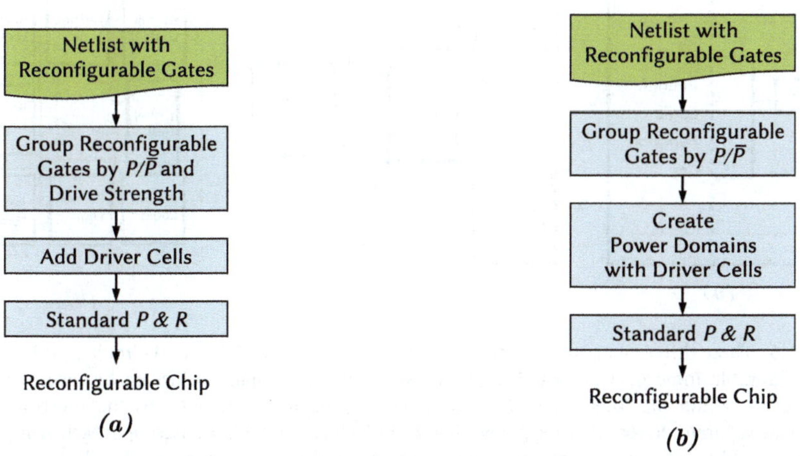

Fig. 6.5 Design flows for (**a**) the native approach and (**b**) the island-based approach

6.7.5 Island-Based Approach

The biggest limitation of the native approach is the unused V_{dd} and V_{ss} contacts on M1, as they only serve for the compatibility of the cells to carry out placements with the default power rails. To circumvent this issue, the connections to V_{dd} and V_{ss} are removed and their metal contacts are reused for P and $\overline{P}$. The improved cell layout is shown in Fig. 6.4c. This approach greatly reduces pin density on M2 in areas with many reconfigurable logic cells. Due to the simpler routing, the total area of the chip and the wirelength are also reduced considerably.

One drawback of this approach is that these new and improved cells can no longer be placed on the V_{dd} and V_{ss} power rails. To still carry out the placement and routing of RFET-based circuits using these improved cell layouts, an island-based approach is applied using multiple power domains and PSO mode, which is supported by Cadence Encounter tool flow for low-power design. In the island-based placement, groups of reconfigurable logic gates that share P and $\overline{P}$ are placed in their own rectangular power domain. As shown in Fig. 6.6c, the driver cells are used as power switches that are automatically placed in the power domain and automatically routed by the *P&R* tool using the PSO mode. Since P and $\overline{P}$ use the power rails in the power domain, no additional resources are needed for routing. By using the switching signal which was originally intended to switch the power domain on and off, the driver cells can be controlled. The polarity of P and $\overline{P}$ can also be reversed depending upon the placement opportunities. This makes it possible to reconfigure all logic gates in the power domain simultaneously, depending on the functionality required by the circuit.

The design flow for this island-based approach is shown in Fig. 6.5b. It starts with the same netlist as the native approach from the previous section. Groups of

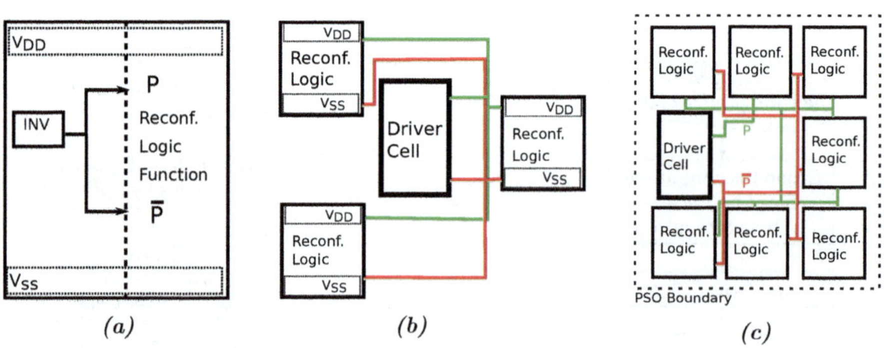

Fig. 6.6 Three different approaches for physical synthesis of RFETs-based circuits. (**a**) Layout of reconfigurable logic gate proposed in [RRK18]. P and $\overline{P}$ signals are driven by inverting logic embedded within the logic gate. (**b**) Our native approach which removes the inverting logic and uses separate driver cells for generating P and $\overline{P}$. (**c**) Our PSO-based approach that groups reconfigurable logic gates in power domains which are driven by common P and $\overline{P}$ signals. (**a**) Standard P&R flow. (**b**) Native flow. (**c**) Island flow

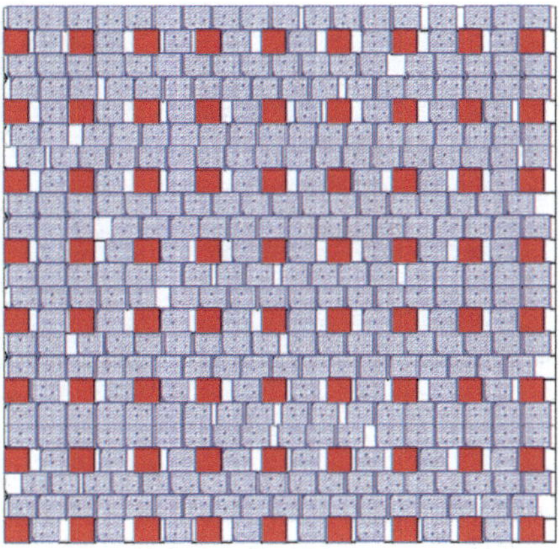

Fig. 6.7 Power domain for reconfigurable gates in an island-based approach that share the same $P/\overline{P}$. Driver cells are highlighted in red. In this example, we added one driver cell for five reconfigurable gates

reconfigurable gates are created that are connected to the same P and $\overline{P}$ signals. This time, however, there is no limit for the size of these groups. For each group, a separate power domain is set up with the appropriate number of driver cells.

As shown in Fig. 6.7, these driver cells are automatically placed in a grid and drive $P/\overline{P}$ using the M1 power rails (cf. Fig. 6.4c). At last, the default standard *P&R* is performed to generate the layout.

6.7.6 Utilization Factor

When placing and routing a netlist, adjusting the utilization factor has the greatest influence on the chances of success and the quality of the layout. This factor defines the ratio between the used and the unused area during placement. For example, a utilization factor of 0.9 means that 90% of the core area is occupied by macros and standard cells, while 10% of the area remains empty. The smaller this factor is, the larger the layout will be, resulting in lower cell and pin densities. As a result, routing gets simplified with this modification.

In the experiments shown in this chapter, *P&R* is started with a core utilization of 0.8 and a power domain utilization of 0.9. Reduced values for these parameters are tried for each benchmark depending on the location of Design Rule Check (DRC) errors until the resulting layout had no DRC violations.

6.7.7 Placement of the Island on the Chip

For the island-based approach, the placement of the power domain in the core is an important consideration. Figure 6.8 shows the three variants that we have examined experimentally. It can be concluded that there is not a single configuration that is

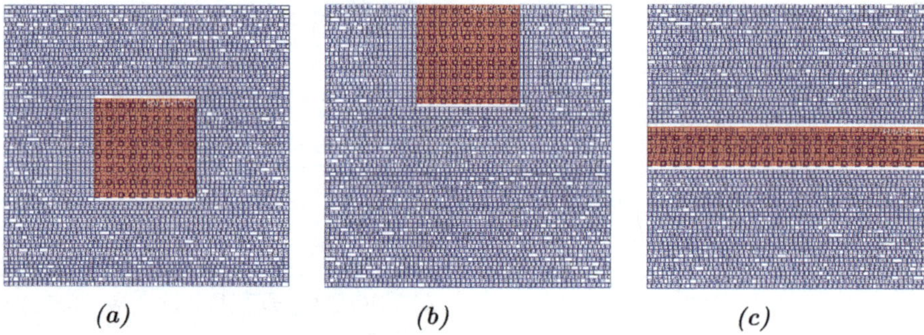

Fig. 6.8 Power domain configurations that we investigated in our experiments: (**a**) square in the center, (**b**) square in the top center, and (**c**) rectangle covering the whole width in the center (stand style)

the best for every test case. Instead, the choice depends on the circuit structure, the amount of reconfigurable logic in a circuit, and whether the optimization criteria is to minimize the area or the wirelength.

As shown in Fig. 6.8, there is an empty row above and below of the power domain. This is necessary because of the different power rail voltages in the power domain and outside. As a result, the stand-style configuration (c) in Fig. 6.8 has a larger unused area than configuration (a). The configuration in (b) has the smallest unused area due to the placement of the power domain at the edge of the circuit. This difference can have a significant influence on the achievable area, especially with smaller circuits. There is no simple answer, as designs react differently to the placement of the island, mainly depending on the circuit structure and the amount of reconfigurable logic a circuit contains.

6.8 Experiments

The experimental evaluations of the physical synthesis for RFETs-based circuits are shown in this section. First, a preliminary comparison with the CMOS technology is shown for both academic and industrial tool flow. This is followed by an exploration of physical synthesis flow where the three proposed approaches are evaluated over EPFL benchmarks [AGD15a].

6.8.1 Preliminary Comparison with CMOS Technology

The table model as explained in the previous section (Sect. 6.6) is used as a base model to generate the `.lef` and `.lib` using the tool flow as shown in Fig. 6.2.

The process of generating these files has been carried out with the following set of gates—*INV, NAND, NOR, XOR, MUX* and *MIN*. Both the open-source (as shown in Fig. 6.3) and the industrial physical synthesis tool flow are carried over MCNC benchmarks [Yan91] using libraries for both SiNW RFETs and the standard CMOS FETs for comparison. The industrial flow is straightforward where we use the netlist generated by ABC and feed it to the Cadence Encounter tool.

For the experiments, a CMOS-based open-source standard cell library is used for comparison. The open-source library from *FreePDK45* [Sti+07] is based on 45-nm CMOS technology. We then use technology scaling as mentioned in [SXB] to have a fair comparison with 22-nm *SOI*-based silicon nanowire library.

Table 6.1 shows the area obtained using both open-source and industrial physical synthesis tool flows. This area is the post place and route area for both CMOS and SiNW technology. From the table, it is clear that SiNW-based circuits occupy more area than the CMOS-based circuit. This can be ascertained with both the industrial and the open-source flow which shows an average overhead of 17.63% and 16.74%, respectively, over the CMOS-based circuits. However, since the experiment uses a 45-nm technology node followed with scaling, it can be expected that at the same

Table 6.1 Preliminary comparison of post-physical synthesis area between CMOS and SiNW RFET technologies over MCNC benchmarks

	QFLOW			Cadence encounter		
Benchmarks	Area CMOS (um2)	Area SiNW (um2)	Increase (%)	Area CMOS (um2)	Area SiNW (um2)	Increase (%)
b1	5.34	5.56	4.10	5.87	5.72	−2.49
b9	49.55	60.60	22.31	51.86	61.20	18.01
C1355	156.17	202.00	29.35	190.39	246.46	29.45
C17	3.41	4.20	23.20	3.20	3.93	22.81
C1908	164.32	202.00	22.93	214.52	262.94	22.57
C432	82.27	89.30	8.54	126.66	151.16	19.34
cm138a	14.07	17.00	20.84	9.73	11.64	19.58
cm150a	13.98	15.90	13.76	28.53	33.98	19.09
cm151a	7.57	8.68	14.69	22.27	26.36	18.38
cm152a	9.18	10.80	17.62	12.40	14.90	20.16
cm162a	19.41	22.70	16.96	21.20	25.06	18.22
cm163a	18.66	22.20	18.98	20.27	24.02	18.51
cm42a	12.91	15.60	20.85	10.27	12.30	19.79
cm82a	9.39	11.20	19.32	11.20	12.96	15.71
cm85a	18.77	21.40	14.00	21.33	25.36	18.87
t481	336.36	401.00	19.22	344.92	419.73	21.69
tcon	18.77	21.00	11.86	24.94	21.38	−14.26
x1	199.55	239.00	19.77	166.39	189.88	14.12
x2	22.95	26.80	16.75	24.80	29.38	18.47
Average			17.63			16.74

technology node, CMOS-based implementation of a given circuit will be smaller by 1–2 units than the overhead observed in the current experiments.

Only a few of the benchmarks from MCNC have been shown here. Practically, physical synthesis for large benchmarks is not possible with the current SiNW library files. This is also observed with the EPFL benchmark suite (used in the next set of experiments) where benchmarks like *hyp* and *div* can only be done with very relaxed constraints of utilization ratio in Cadence Encounter. On the other hand, the open-source flow is not even able to perform the P&R for large circuits. In fact, two extra metal layers were also added in the SiNW `.lef` file to ease the routing constraints. This was primarily done to tackle the high pin density in the case of layouts of SiNW-based standard cells. Apart from that, the *design rule check* (DRC) in the open-source flow is not so robust as compared to the industrial tool flow. Hence, there is a bit of discrepancy in the area of the individual benchmarks for the two flows. For the industrial flow, a utilization ratio of 0.8 is used for the SiNW and CMOS technology for all the benchmarks. Only for *cm82a*, a utilization factor of 0.7 is used for the SiNW technology. This setting of utilization ratio ensures no DRC violations. It can be seen that both the flows achieve a similar average result of 17.64% and 16.74% for the open-source and industrial flow, respectively.

6.8.2 Evaluating Different Physical Synthesis Approaches

In this section, experiments with the two new approaches are presented for optimized physical synthesis for emerging reconfigurable nanotechnologies. For the experiments, the same silicon nanowire-based RFET model is used. The *P&R* is carried out using Cadence Encounter. The baseline flow is the default standard P&R flow using the RFET model as shown with the industrial flow described in Sect. 6.8.1. This is compared with the native and island-based approaches.

For the sake of simplicity, a trivial example of logic locking is taken in which the starting RFET-based circuit is only mapped with *NAND* and *NOR* logic gates. Then, some portions of these gates are replaced with *MINORITY* gate (a polymorphic gate which can be configured either as *NAND* or *NOR*) to carry out the physical synthesis flow to demonstrate our approach and findings. This has been done to show a practical use case for RFETs-based reconfigurable logic gates in circuits for security applications. This is also driven by the limitations of not having multigate RFETs to construct large logic gates such as the 3-XOR or the reconfigurable 3-NAND_3-NOR.

The three approaches (baseline, native, and island-based approaches) are applied to the EPFL benchmark suite [AGD15b]. First, these circuits are mapped using the ABC logic synthesis tool [BM10] and a target library containing only *NAND* and *NOR* gates. To create reconfigurable circuits for the two approaches (native and island-based approaches), a part of these *NAND* and *NOR* gates are replaced with reconfigurable *MIN* gates (cf. Fig. 6.1c). These *MIN* gates represent the reconfigurable logic gates used for hardware security applications [Rai+20b]. While the default P&R flow uses the *MIN* gate layouts from [RRK18] shown in Fig. 6.4a, the netlists of both the native approach and the island-based approach use the new layouts shown in Fig. 6.4b and c, respectively.

The *NAND* and *NOR* gates to be replaced are selected evenly across the entire netlist. This is modeled using a parameter called as *replacement ratio*. For this purpose, all candidate gates are iterated over and a gate is only selected if the current replacement ratio is below the desired replacement ratio. Since reconfigurable gates are placed in separate power domains during the island-based approach, this selection requires an additional optimization step. For this, the Fiduccia-Mattheyses (FM) algorithm [FM82] is applied to reduce the number of cut nets between the reconfigurable logic gates and the rest of the circuit. Starting with the initial partitioning, two steps are performed repeatedly: (1) move a gate *out* of the power domain, and (2) move another gate *into* the power domain. By combining the two steps, the number of gates in the power domain and, therefore, the replacement ratio remain constant. The gates are selected in the order of their gain, i.e., the more a gate reduces the number of cut nets, the sooner it is moved. Only if there is an improvement, the two moves are executed. After all gates in the power domain have been considered and as long as the number of cut nets has been reduced, the whole process starts again.

The entire process of

1. Partitioning and transforming the netlist
2. Template-based creation of the TCL scripts and CPF files to control Cadence Encounter
3. Running *P&R*
4. Creating the layout images (cf. Fig. 6.9)
5. Extracting the relevant parameters from the log files

is automated using Python.

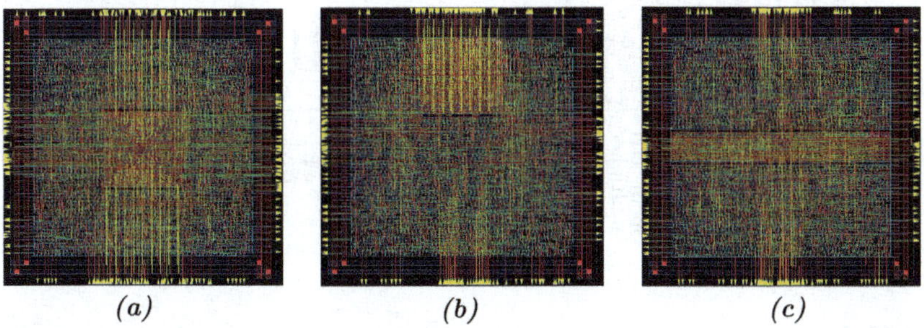

Fig. 6.9 Layouts of the EPFL *bar* benchmark with 10% reconfigurable logic using the island-based approach. The three layouts correspond to the floorplans in Fig. 6.8

6.9 Results and Discussions

Figure 6.10 shows the results of our experiment for a replacement ratio of 0.1. Both approaches described in Sect. 6.7 (native and island-based approach) are shown in comparison with the baseline. As we can see in Fig. 6.10a, the core area is reduced by 1.5% on average (2.3% when ignoring the outlier benchmark *div*) for the native approach and by 0.2% for the island-based approach. These values are quite low for

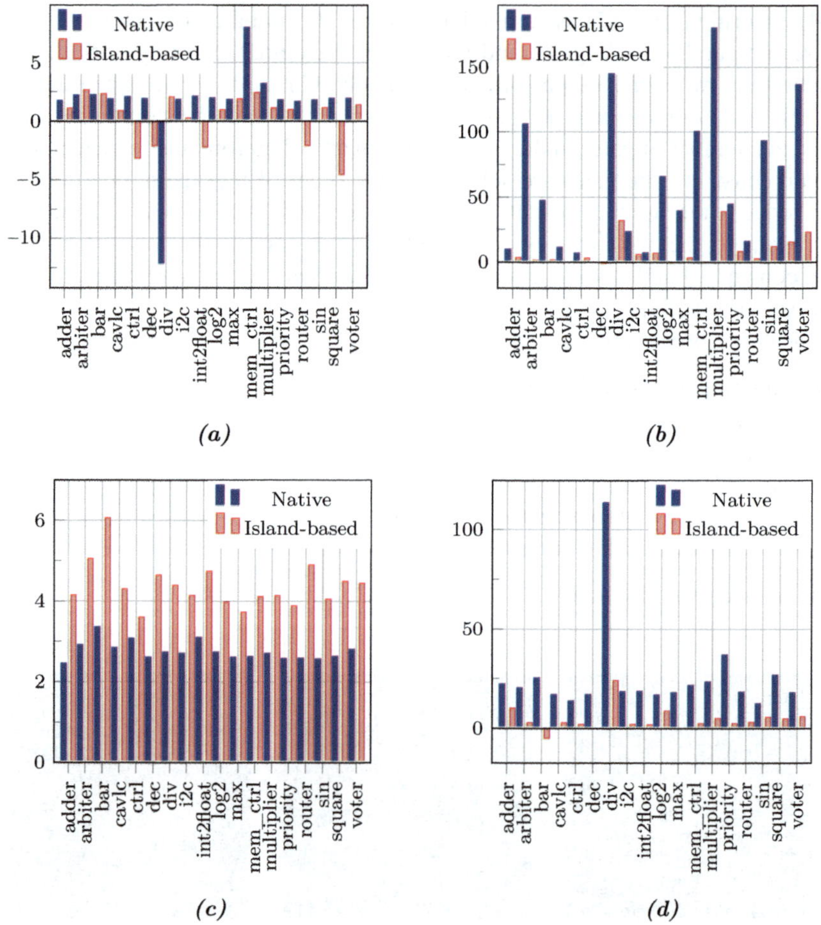

Fig. 6.10 Comparison of the native approach and the island-based approach with the baseline for a replacement ratio of 0.1, i.e., 10% of the logic gates are replaced by reconfigurable logic gates. For each EPFL benchmark, the changes in (**a**) core area, (**b**) HPWL, (**c**) static power consumption, and (**d**) dynamic power consumption from the baseline are shown. For the island-based approach we use the best results of the three power domain configurations in Fig. 6.8. Please note that in (**a**) and (**c**) higher values are better, while in (**b**) and (**d**) lower values are desired. (**a**) Area reduction (%). (**b**) HPWL increase (%). (**c**) Static power reduction (%). (**d**) Dynamic power increase (%)

two reasons. First, a large part of the area advantage is lost—since the inverter is removed from the reconfigurable logic gates—by adding one driver cell for each group of five[2] reconfigurable logic gates. This shows that the number of gates a driver cell can drive has a great impact on the area. Second, the island-based approach has a high overhead due to the empty rows above and below, which makes the area reduction even worse on average compared to the native approach.

However, as shown in Fig. 6.10b, the island-based approach has a much smaller impact on HPWL. This is caused by the high pin density when using the native approach because each reconfigurable logic gate has two additional pins, $P/\overline{P}$. With the island-based approach, these pins are connected using the power rails, which reduces the pin density and, thereby, HPWL. This high pin density of the standard cell layout is because the original layouts of the logic gates are very CMOS-like, hence ignoring the structural differences with the RFETs. This does not allow easy routing for metal layers M2 within the island. As a result, the P and $\overline{P}$ routing is done either through metal layers M3 or M4 in order to make the physical synthesis DRC-free. This can be, however, mitigated with better standard cell designs using a more advanced PDK. Specially RFET-based PDK can alleviate these issues and can also enable further area reduction with better standard cell designs.

The static power consumption as shown in Fig. 6.10c, as reported by Cadence Encounter, is reduced in both approaches. In contrast, the dynamic power consumption as shown in Fig. 6.10d is increased for most of the benchmarks. This increase is far worse for the native approach, where it correlates with the higher HPWL.

When looking at the influence of the power domain configuration discussed in Sect. 6.7.7, a square in the top center (Fig. 6.8b) minimizes the core area for all benchmark circuits. For HPWL, a square power domain at the top or in the center is generally the best (Fig. 6.8a or b), except for some very large circuits (benchmarks *voter*, *square*, *log2*) where the configuration in Fig. 6.8c (stand-style) produces the layout with the smallest HPWL.

In the second experiment, both approaches are applied to the benchmark *bar* using a replacement ratio from 0.1 to 1.0. The results are shown in Fig. 6.11. As we can see, the island-based approach is far superior for larger replacement ratios. The area overhead decreases with more reconfigurable logic gates and larger power domains. As a result, the area is reduced by up to 17.5% for a replacement ratio of 0.9 (Fig. 6.11a). Again, the wirelength increase is much lower for the island-based approach, with a maximum of 13.0% for a replacement ratio of 0.6 (Fig. 6.11b). The static power reduction in Fig. 6.11c is proportional to the replacement ratio. The change in dynamic power consumption varies between a decrease of 13.1% and an increase of 5.9% (Fig. 6.11d).

For the island-based approach, the smallest area is achieved using the floorplan in Fig. 6.8b with the power domain centered at the top. The optimal configuration for the smallest HPWL depends on the replacement ratio. For a replacement ratio of

[2] The driver cell can be designed with any drive strength. Here, the experiments have been carried out with the drive strength assumed to be 5.

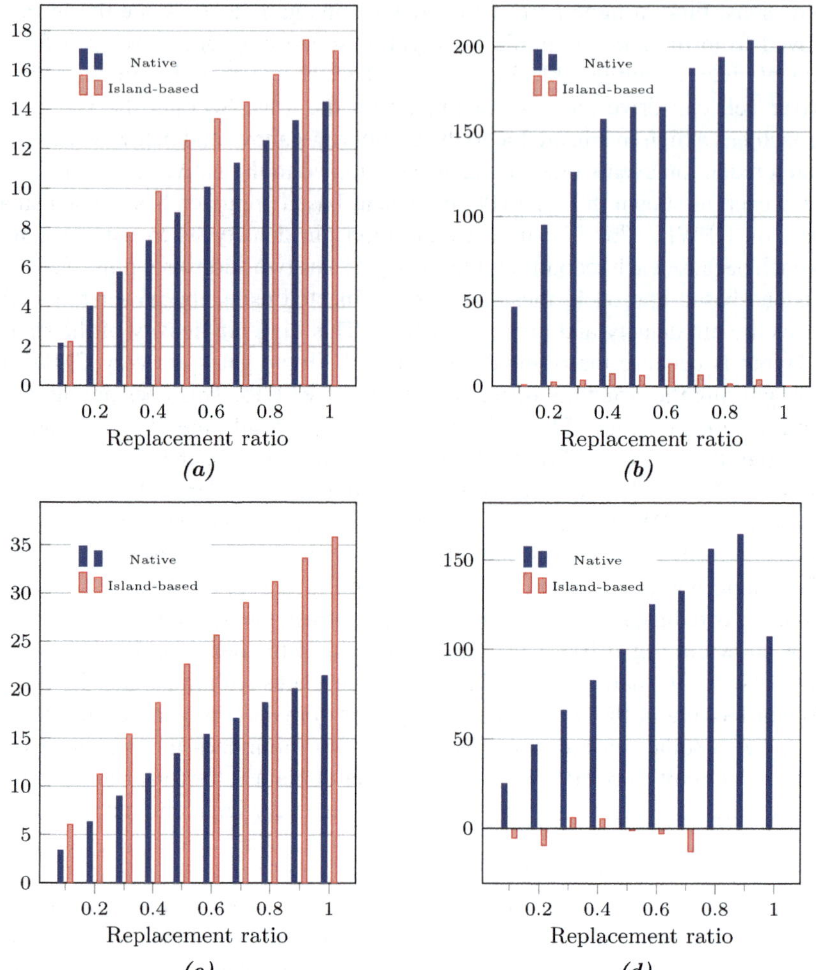

Fig. 6.11 Changes in (**a**) core area, (**b**) HPWL, (**c**) static power consumption, and (**d**) dynamic power consumption from the baseline for the EPFL *bar* benchmark with the amount of reconfigurable logic ranging from 10 to 100%. For the island-based approach we used the best results of the three power domain configurations in Fig. 6.8. Please note that in (**a**) and (**c**) higher values are better, while in (**b**) and (**d**) lower values are desired. (**a**) Area reduction (%). (**b**) HPWL increase (%). (**c**) Static power reduction (%). (**d**) Dynamic power increase (%)

0.1 or 0.2, the floorplan in Fig. 6.8b produces the smallest HPWL. For a replacement ratio between 0.3 and 0.5, the centered square in Fig. 6.8a is the best configuration with regard to HPWL. If more than 50% of all gates are reconfigurable, the stand-style configuration in Fig. 6.8c is the best choice.

From our exploration study, we can see that there are a lot of parameters that affect the physical synthesis of the circuits based on RFETs. One can see that the island-based approach is more suitable and gives consistently better results in terms of area and wirelength as compared to the native-based approach. The following findings can be noted from our experiments:

- All three approaches—baseline, native, or island-based approach—can be used depending upon the parameter to optimize. For example, the native flow suffers through the HPWL overhead but often gives better area reduction as compared to the baseline flow.
- The factor of replacement ratio which is basically governed by the security assurance has a huge impact in terms of area and wirelength.
- Placement of the power domains depends on factors such as the size of the circuit, connections between the subcircuits, and also the layout of the standard cells. Learning-based approaches seem suitable to effectively predict optimized power domain placement. Alternatively, top-level power domain position requirements and gate-level security requirements can be translated into constraints that can be efficiently communicated through top-down and bottom-up propagation within the design hierarchy [KJL15]. These propagated constraints can then support informed configuration of the power domains [KLL17].
- The quality of *liberty* and *lef* files plays an important role in placement and routing. Factors such as pitch between the gate terminals and pin density need to be optimized for shorter total routed wirelength.

6.9.1 Parameters Which Affect the Area

The present chapter presents a complete feasibility study of the design route by taking into account a simplified technology flow for the gate patterning and manufacturing. Nevertheless, as stated in Sect. 6.4, it is technically possible to substantially reduce the length of the transistor cells by the use of self-aligned techniques for double-gate patterning in combination with an additional lithography step. This additional technological effort would bring a benefit in the circuit size. The RFET could be realized with a total channel length of 48 nm, compared to the 100 nm used here. The electrical performance of such a scaled RFET device has already been verified by TCAD simulation and benchmarked in terms of inverter delay and power consumption in the work of [Tro+15]. The transistor cell length could be reduced from a total length of $\sim$188 nm down to 92 nm. Consequently, the layouts of Fig. 6.1 will substantially reduce in size, although not by the same factor as the design rules have to be met. Another opportunity to actively reduce the device area is vertical stacking of multiple nanowires ([Mer+16, Ern+06]) as already demonstrated for polarity control devices with four stacked nanowires by De Marchi et al. [Mar+12]. This will lead to a substantial reduction of device width and accordingly to a reduced cell height of the layouts in Fig. 6.1. One can anticipate that reduction in nanowire channel length will have a positive impact in the reduction of capacitances and RC delay of the transistor [Tro+15]. For stacked nanowires, methods on controlling RC delays explored in [Ber+09] can be applied to RFET devices as well.

6.9.2 *Use of Germanium Nanowire Channels*

In terms of circuit speed and power consumption, a promising solution is to use germanium or silicon-germanium nanowire channel instead of silicon. Silicon germanium and pure germanium channels are conventionally used for p-FETs. Indeed, it has been shown experimentally and by TCAD simulations [Tro+17b] that the use of germanium nanowires for RFETs is feasible. Germanium brings benefits in the performance and power consumption of RFETs both for p-type and n-type characteristics. The lower bandgap of germanium together with the higher tunneling probability is able to enhance the device drive currents by a factor of 10. The expense for taking this route is a higher static current. However, the inherent blocking nature of the RFET's program gate is able to filter out a substantial level of source-drain leakage. The lower Ge bandgap also enables the reduction of the supply voltage of RFETs from 1.8 V to $\sim$ 0.8 V, thereby reducing dynamic power consumption of the transistors to less than one-fourth of the SiNW RFET value.

6.10 Concluding Remarks

In this chapter, a physical synthesis flow for technology evaluation for silicon nanowire reconfigurable FETs has been presented. RFETs offer a new paradigm as compared to CMOS because of their reconfigurable properties where logic gates can be either reconfigurable or static. Such reconfigurable components often constitute the control segment of the circuit as such features are highly suited for applications in hardware security. In this direction, both static and reconfigurable layout designs have been presented for a few of the logic gates based on RFETs. The present layout designs are inspired by CMOS layout designs and that is why there are certain limitations in the physical synthesis flow. An extensive evaluation of the physical synthesis flow for MCNC benchmarks in terms of area is carried out using both academic and industrial tool flows. It has been observed that both the flows achieve a similar result with an average overhead of 17.63% (open-source tool flow) and 16.74% (industrial flow) for the SiNW-based circuits over the CMOS-based circuits.

Keeping security application in mind, two strategies for physical synthesis have been explored with different placement techniques to achieve further area reductions by dedicating separate areas for reconfigurable logic gates and static logic gates in the circuit. Using a PSO-inspired island-based approach, improvements in physical parameters such as static power consumption and area have been shown. While most benchmark circuits perform well when the island is a square in the center of the die, a detailed study has to be carried out in order to find the underlying reasons why a particular placement works for a circuit.

References

[AGD15a] Luca Amarú, Pierre-Emmanuel Gaillardon, and Giovanni De Micheli. "The EPFL combinational benchmark suite". In: *Proceedings of the 24th International Workshop on Logic & Synthesis (IWLS)*. 2015.

[AGD15b] Luca Amarú, Pierre-Emmanuel Gaillardon, and Giovanni De Micheli. "The EPFL combinational benchmark suite". In: *Proceedings of the 24th International Workshop on Logic & Synthesis (IWLS)*. CONF. 2015.

[AGM13] L. Amarù, P. E. Gaillardon, and G. De Micheli. "Efficient arithmetic logic gates using double-gate silicon nanowire FETs". In: *2013 IEEE 11th International New Circuits and Systems Conference (NEWCAS)*. June 2013, pp. 1–4. https://doi.org/10.1109/NEWCAS.2013.6573572.

[AGM16] L. Amarú, P. E. Gaillardon, and G. De Micheli. "Majority-Inverter Graph: A New Paradigm for Logic Optimization". In: *IEEE Transactions on Computer-Aided Design of Integrated Circuits and Systems* 35.5 (May 2016), pp. 806–819. https://doi.org/10.1109/TCAD.2015.2488484.

[Bal+17] Tim Baldauf, André Heinzig, Jens Trommer, Thomas Mikolajick, and Walter Michael Weber. "Tuning the tunneling probability by mechanical stress in Schottky barrier based reconfigurable nanowire transistors". In: *Solid-State Electronics* 128 (2017). Extended papers selected from EUROSOI-ULIS 2016, pp. 148–154. https://doi.org/10.1016/j.sse.2016.10.009.

[Ber+09] E. Bernard, T. Ernst, B. Guillaumot, N. Vulliet, X. Garros, P. Coronel, T. Skotnicki, S. Deleonibus, and O. Faynot. "Multi-Channel Field-Effect Transistor (MCFET) 2014;Part II: Analysis of Gate Stack and Series Resistance Influence on the MCFET Performance". In: *IEEE Transactions on Electron Devices* 56.6 (June 2009), pp. 1252–1261. https://doi.org/10.1109/TED.2009.2019155.

[Bi+16] Yu Bi, Kaveh Shamsi, Jiann-Shiun Yuan, Pierre-Emmanuel Gaillardon, Giovanni De Micheli, Xunzhao Yin, X Sharon Hu, Michael Niemier, and Yier Jin. "Emerging technology-based design of primitives for hardware security". In: *ACM Journal on Emerging Technologies in Computing Systems (JETC)* 13.1 (2016), pp. 1–19.

[BM10] Robert Brayton and Alan Mishchenko. "ABC: An academic industrialstrength verification tool". In: *Computer Aided Verification*. Springer. 2010.

[BM15] S. Bobba and G. De Micheli. "Layout Technique for Double-Gate Silicon Nanowire FETs With an Efficient Sea-of-Tiles Architecture". In: *IEEE Transactions on Very Large Scale Integration (VLSI) Systems* 23.10 (Oct. 2015), pp. 2103–2115. https://doi.org/10.1109/TVLSI.2014.2358884.

[Bob+12] Shashikanth Bobba, Michele De Marchi, Yusuf Leblebici, and Giovanni De Micheli. "Physical Synthesis Onto a Sea-of-Tiles with Double-gate Silicon Nanowire Transistors". In: *Proceedings of the 49th Annual Design Automation Conference*. DAC '12. San Francisco, California: ACM, 2012, pp. 42–47. https://doi.org/10.1145/2228360.2228369.

[Cha+20] A. Chakraborty, N. G. Jayasankaran, Y. Liu, J. Rajendran, O. Sinanoglu, A. Srivastava, Y. Xie, M. Yasin, and M. Zuzak. "Keynote: A Disquisition on Logic Locking". In: *IEEE Transactions on Computer- Aided Design of Integrated Circuits and Systems* 39.10 (2020), pp. 1952–1972. https://doi.org/10.1109/TCAD.2019.2944586.

[Edw17] Tim Edwards. *A Digital Synthesis Flow using Open Source EDA Tools*. 2017. url: http://opencircuitdesign.com/qflow/index.html.

[Ern+06] T. Ernst, C. Dupre, C. Isheden, E. Bernard, R. Ritzenthaler, V. Maffini-Alvaro, J. C. Barbe, F. De Crecy, A. Toffoli, C. Vizioz, S. Borel, F. Andrieu, V. Delaye, D. Lafond, G. Rabille, J. M. Hartmann, M. Rivoire, B. Guillaumot, A. Suhm, P. Rivallin, O. Faynot, G. Ghibaudo, and S. Deleonibus. “Novel 3D integration process for highly scalable Nano-Beam stacked-channels GAA (NBG) FinFETs with HfO2/TiN gate stack”. In: *2006 International Electron Devices Meeting*. Dec. 2006, pp. 1–4. https://doi.org/10.1109/IEDM.2006.346955.

[FM82] C. M. Fiduccia and R. M. Mattheyses. “A Linear-Time Heuristic for Improving Network Partitions”. In: *DAC*. 1982.

[Gai+13] Pierre-Emmanuel Gaillardon, Luca Gaetano Amarù, Shashikanth Bobba, Michele De Marchi, Davide Sacchetto, Yusuf Leblebici, and Giovanni De Micheli. “Vertically-stacked Double-gate Nanowire FETs with Controllable Polarity: From Devices to Regular ASICs”. In: *Proceedings of the Conference on Design, Automation and Test in Europe*. DATE ’13. Grenoble, France: EDA Consortium, 2013, pp. 625–630.

[Gai+14] Pierre Emmanuel Gaillardon, Shashikanth Bobba, Michele De Marchi, Davide Sacchetto, and Giovanni De Micheli. “Nanowire systems: technology and design”. In: *Philosophical Transactions of the Royal Society of London* (2014).

[Gor+19] G. Gore, P. Cadareanu, E. Giacomin, and P. Gaillardon. “A Predictive Process Design Kit for Three-Independent-Gate Field-Effect Transistors”. In: *2019 IFIP/IEEE 27th International Conference on Very Large Scale Integration (VLSI-SoC)*. Oct. 2019, pp. 172–177. https://doi.org/10.1109/VLSI-SoC.2019.8920358.

[Gra17] Graywolf. *Graywolf—Professional-grade placement tool*. 2017. url: https://github.com/rubund/graywolf.

[Hei+12] André Heinzig, Stefan Slesazeck, Franz Kreupl, Thomas Mikolajick, and Walter M. Weber. “Reconfigurable silicon nanowire transistors”. In: *Nano Letters* 12.1 (2012). PMID: 22111808, pp. 119–124. https://doi.org/10.1021/nl203094h. eprint: https://doi.org/10.1021/nl203094h.

[KJL15] Andreas Krinke, Goeran Jerke, and Jens Lienig. “Constraint Propagation Methods for Robust IC Design”. In: 2015, pp. 7–14.

[KLL17] Andreas Krinke, Lei Lei, and Jens Lienig. “Predictive System-Level Constraint Verification and Optimization”. In: 2017, pp. 40–45.

[Kne20] Johann Knechtel. “Hardware Security For and Beyond CMOS Technology: An Overview on Fundamentals, Applications, and Challenges”. In: *Proceedings of the 2020 International Symposium on Physical Design*. New York, NY, USA: Association for Computing Machinery, 2020, pp. 75–86.

[Köf17] Matthias Köfferlein. *KLayout—High Performance Layout Viewer And Editor*. 2017. url: http://www.klayout.de/index.html.

[Mac11] C. A. Mack. “Fifty Years of Moore’s Law”. In: *IEEE Transactions on Semiconductor Manufacturing* 24.2 (May 2011), pp. 202–207. https://doi.org/10.1109/TSM.2010.2096437.

[Mar+12] M. De Marchi, D. Sacchetto, S. Frache, J. Zhang, P. E. Gaillardon, Y. Leblebici, and G. De Micheli. “Polarity control in double-gate, gate-all-around vertically stacked silicon nanowire FETs”. In: *2012 International Electron Devices Meeting*. Dec. 2012, pp. 8.4.1–8.4.4. https://doi.org/10.1109/IEDM.2012.6479004.

[Mer+16] H. Mertens, R. Ritzenthaler, A. Chasin, T. Schram, E. Kunnen, A. Hikavyy, L. Å. Ragnarsson, H. Dekkers, T. Hopf, K. Wostyn, K. Devriendt, S. A. Chew, M. S. Kim, Y. Kikuchi, E. Rosseel, G. Mannaert, S. Kubicek, S. Demuynck, A. Dangol, N. Bosman, J. Geypen, P. Carolan, H. Bender, K. Barla, N. Horiguchi, and D. Mocuta. “Vertically stacked gate-all-around Si nanowire CMOS transistors with dual work function metal gates”. In: *2016 IEEE International Electron Devices Meeting (IEDM)*. Dec. 2016, pp. 19.7.1–19.7.4. https://doi.org/10.1109/IEDM.2016.7838456.

[Qro17] Qrouter. *Qrouter over the cell router or sea of gates detailed router*. 2017. url: http://opencircuitdesign.com/qrouter/.

[Rai+17] M. Raitza, A. Kumar, M. Völp, D.Walter, J. Trommer, T. Mikolajick, and W. M. Weber. "Exploiting transistor-level reconfiguration to optimize combinational circuits". In: *Design, Automation Test in Europe Conference Exhibition (DATE), 2017*. Mar. 2017, pp. 338–343. https://doi.org/10.23919/DATE.2017.7927013.

[Rai+18] S. Rai, A. Rupani, D. Walter, M. Raitza, A. Heinzig, T. Baldauf, J. Trommer, C. Mayr, W. M. Weber, and A. Kumar. "A physical synthesis flow for early technology evaluation of silicon nanowire based reconfigurable FETs". In: *2018 Design, Automation Test in Europe Conference Exhibition (DATE)*. Mar. 2018, pp. 605–608. https://doi.org/10.23919/DATE.2018.8342080.

[Rai+19a] S. Rai, A. Rupani, P. Nath, and A. Kumar. "Hardware Watermarking Using Polymorphic Inverter Designs Based On Reconfigurable Nanotechnologies". In: *ISVLSI*. 2019.

[Rai+19b] S. Rai, J. Trommer, M. Raitza, T. Mikolajick, W. M. Weber, and A. Kumar. "Designing Efficient Circuits Based on Runtime-Reconfigurable Field-Effect Transistors". In: *IEEE Transactions on Very Large Scale Integration (VLSI) Systems* 27.3 (Mar. 2019), pp. 560–572. https://doi.org/10.1109/TVLSI.2018.2884646.

[Rai+20a] S. Rai, M. Raitza, S.S. Sahoo, and A. Kumar. "DiSCERN: Distilling Standard-Cells for Emerging Reconfigurable Nanotechnologies". In: *DATE*. 2020.

[Rai+20b] Shubham Rai, Satwik Patnaik, Ansh Rupani, Johann Knechtel, Ozgur Sinanoglu, and Akash Kumar. "Security Promises and Vulnerabilities in Emerging Reconfigurable Nanotechnology-Based Circuits". In: *IEEE Transactions on Emerging Topics in Computing* (2020), pp. 1–1. https://doi.org/10.1109/TETC.2020.3039375.

[Reu+21] Maximilian Reuter, Johannes Pfau, Tillmann A. Krauss, Jürgen Becker, and Klaus Hofmann. "From MOSFETs to Ambipolar Transistors: Standard Cell Synthesis for the Planar RFET Technology". In: *IEEE Transactions on Circuits and Systems I: Regular Papers* 68.1 (2021), pp. 114–125. https://doi.org/10.1109/TCSI.2020.3035889.

[RRK18] S. Rai, M. Raitza, and A. Kumar. "Technology mapping flow for emerging reconfigurable silicon nanowire transistors". In: *2018 Design, Automation Test in Europe Conference Exhibition (DATE)*. Mar. 2018, pp. 767–772. https://doi.org/10.23919/DATE.2018.8342110.

[Sin+21] Ozgur Sinanoglu, Shaloo Rakheja, Johann Knechtel, Satwik Patnaik, and Nikhil Rangarajan. *The Next Era in Hardware Security: A Perspective on Emerging Technologies for Secure Electronics*. Springer, 2021.

[SRM15] P. Subramanyan, S. Ray, and S. Malik. "Evaluating the security of logic encryption algorithms". In: *Proc. HOST*. 2015, pp. 137–143. https://doi.org/10.1109/HST.2015.7140252.

[Sti+07] J. E. Stine, I. Castellanos, M. Wood, J. Henson, F. Love, W. R. Davis, P. D. Franzon, M. Bucher, S. Basavarajaiah, J. Oh, and R. Jenkal. "FreePDK: An Open-Source Variation-Aware Design Kit". In: *2007 IEEE International Conference on Microelectronic Systems Education (MSE'07)*. June 2007, pp. 173–174. https://doi.org/10.1109/MSE.2007.44.

[SXB] Aaron Stillmaker, Zhibin Xiao, and Bevan Baas. "Toward more accurate scaling estimates of cmos circuits from 180 nm to 22 nm". In: *Technical Report ECE VCL 2011 4 VLSI Computation Lab, University of California, Davis* ().

[Tro+15] Jens Trommer, André Heinzig, Tim Baldauf, Stefan Slesazeck, Thomas Mikolajick, and Walter M. Weber. "Functionality-Enhanced Logic Gate Design Enabled by Symmetrical Reconfigurable Silicon Nanowire Transistors". In: *IEEE Transactions on Nanotechnology* 14.4 (July 2015), pp. 689–698. https://doi.org/10.1109/TNANO.2015.2429893.

[Tro+16] J. Trommer, A. Heinzig, T. Baldauf, T. Mikolajick, W. M. Weber, M. Raitza, and M. Völp. "Reconfigurable nanowire transistors with multiple independent gates for efficient and programmable combinational circuits". In: *2016 Design, Automation Test in Europe Conference Exhibition (DATE)*. Mar. 2016, pp. 169–174.

[Tro+17a] J Trommer, A Heinzig, S Slesazeck, U Mühle, M Löffler, D Walter, C Mayr, T Mikolajick, and WM Weber. "Reconfigurable germanium transistors with low source-drain leakage for secure and energyefficient doping-free complementary circuits". In: *Device Research Conference (DRC), 2017 75th Annual*. IEEE. 2017, pp. 1–2.

[Tro+17b] Jens Trommer, André Heinzig, Uwe Mühle, Markus Löffler, Annett Winzer, Paul M. Jordan, Jürgen Beister, Tim Baldauf, Marion Geidel, Barbara Adolphi, Ehrenfried Zschech, Thomas Mikolajick, and Walter M. Weber. "Enabling Energy Efficiency and Polarity Control in Germanium Nanowire Transistors by Individually Gated Nanojunctions". In: *ACS Nano* 11.2 (2017). PMID: 28080025, pp. 1704–1711. https://doi.org/10.1021/acsnano.6b07531. eprint: https://dx.doi.org/10.1021/acsnano.6b07531.

[Wol] CliffordWolf. *Yosys Open SYnthesis Suite*. http://www.clifford.at/yosys/.

[Yan91] Saeyang Yang. *Logic Synthesis and Optimization Benchmarks User Guide Version* 3.0. 1991.

[ZGD13] Jian Zhang, Pierre Emmanuel Gaillardon, and Giovanni De Micheli. "Dual-threshold-voltage configurable circuits with three-independentgate silicon nanowire FETs". In: *Proceedings—IEEE International Symposium on Circuits and Systems*. May 2013, pp. 2111–2114. https://doi.org/10.1109/ISCAS.2013.6572291.

[Zha+14] Jian Zhang, Xifan Tang, Pierre Emmanuel Gaillardon, and Giovanni De Micheli. "Configurable Circuits Featuring Dual-Threshold-Voltage Design With Three-Independent-Gate Silicon Nanowire FETs". In: *IEEE Transactions on Circuits and Systems I:Regular Papers* 61.10 (Oct. 2014), pp. 2851–2861. https://doi.org/10.1109/TCSI.2014.2333675.

Chapter 7
Polymorphic Primitives for Hardware Security

While the previous chapters focused on enabling design automation techniques for Reconfigurable Field-Effect Transistors (RFETs)-based circuits, this chapter explores hardware security as an application using RFETs-based primitives. This chapter presents efficient and affordable Intellectual Property (IP) protection measures that can be realized using the reconfigurable properties of RFETs.

In the last decade, hardware security has become an important concern for circuit designers along with the metrics of power, area, and delay. It is imperative for circuit designers to consider security implications across different stages of chip development to ensure that any kind of adversarial attacks can be thwarted. The prime reason can be ascertained by the globalization of the semiconductor industry. Globalization of the electronic industry supply chain spread across various continents makes it easier for adversaries to device new and vicious attacks. This has led to frequent copyright infringements and illegal ownership claims on IP. According to a report [SEM12], the Integrated Circuits (IC) industry, on average, loses about $4 billion annually due to design infringements. As a result, fake ICs or consumer products using these fake ICs are common. This causes huge monetary losses to the original IP designer.

Hardware security threats can manifest themselves in two major ways—firstly, the adversary can embed certain malicious components within the die of the circuit that can leak privacy-related information of the user or can even lead to compromise in Quality of Service (QoS) and other performance-related issues in the host circuit. Secondly, if the circuit is not secured or *obfuscated* enough, a counterfeit or a duplicate copy of the circuit can be produced which can hamper the sale or image of a particular IC.

In recent years, there has been a great thrust in devising various security measures across the Electronic Design Automation (EDA) stack. This chapter particularly focuses on enabling hardware security features from the device level. RFET-based circuits possess security-rich features owing to the device-level dynamic reconfiguration of electrical properties. Ambipolarity in RFETs enables various

S. Rai, A. Kumar, *Design Automation and Applications for Emerging Reconfigurable Nanotechnologies*, https://doi.org/10.1007/978-3-031-37924-6_7

efficient and cost-saving security features that are available inherently in RFETs-based circuits.

Before getting into the details of how RFETs enable hardware security, it is necessary to first understand what *hardware security* is and how it is different compared to *cyber security*. Hardware security deals with the integrity of the underlying hardware, while cyber security primarily deals with the security of the software solutions running on that hardware. While the two terms are somewhat orthogonal to each other, recent attacks such as *Meltdown* [Lip+18], *Spectre* [Koc+18], and *Rowhammer* [Kim+14] have blurred the gap between the two. The scope of this chapter is however restricted to hardware security. Both security benefits (such as logic locking, split manufacturing) and vulnerability in RFETs-based circuits are discussed in this chapter. The chapter specifically targets the following research question:

Research Question How are polymorphic logic gates designed by utilizing device-level reconfigurability offered by RFETs to develop efficient and secure circuits? Does the security offered by RFETs strong enough to tackle the state-of-the-art security techniques? And is the security the only gain, or are the RFETs-based circuits have certain vulnerabilities? With the above research questions in mind, the chapter discusses what the major challenges and opportunities are that need to be explored for enabling security for circuits using RFETs.

7.1 Contributions

The primary contributions of this chapter are as follows:

1. Functional polymorphism in logic gates based on RFETs is demonstrated. The functional polymorphism is leveraged for IP protection schemes such as logic locking and split manufacturing.
2. Security analysis is provided for the proposed IP protection schemes to demonstrate efficient practical security for circuits based on RFETs. Experiments using state-of-the-art attack schemes demonstrate security effectiveness.
3. The same functional polymorphism can be detrimental making circuits based on RFETs vulnerable to security attacks. This is realized by misconfiguration of one or more transistors to disrupt the current normal functionality.
4. Evaluations show that the abovementioned security vulnerabilities have an adverse effect on current and voltage levels for both combinational and sequential circuits.

7.2 Organization

In this chapter, we will first look in Sect. 7.3 at why and how the research community started looking at solutions based on emerging or beyond Complementary Metal

Oxide Semiconductor (CMOS) technologies to provide security guarantees. This is followed by the introduction of various IP protection schemes realizable using RFETs in Sect. 7.4. After this background about these security schemes, measures that have been specifically proposed using RFETs are then discussed in Sect. 7.5. Then, scenarios related to the security vulnerability in RFETs-based circuits are presented in Sect. 7.6. We study two important scenarios—short circuit and open circuit for RFET-based inverters—and then explore such scenarios that can be realized in combinational and sequential circuits. This is followed by extensive evaluations for both security benefits and vulnerabilities in Sect. 7.7. Finally, the concluding remarks and takeaways for future research directions are presented in Sect. 7.8.

7.3 The Shift to Explore Emerging Technologies for Security

An important aspect to harness circuit-level hardware security is to exploit functional *polymorphism* at the logic gate or at the circuit level. Polymorphism refers to a property by which a single block of hardware can show multiple logical functionalities. A hardware block or a unit demonstrating a range of functionalities is able to thwart adversarial attacks that are primarily aimed at deciphering the logic-in-use. In [McD+16], the concept of polymorphic gates has been demonstrated for CMOS, where various gates in a netlist are randomly chosen and replaced by polymorphic gates, which are activated for a particular function only when a predefined key is provided. Another form of a polymorphic logic gate has been defined in [SR14]. In this work, Simek et al. have devised a polymorphic gate that can behave as a NAND gate or a NOR gate depending on the applied voltage potential V_{DD}. However, these polymorphic solutions often lead to huge overheads in terms of area and power.

In [Che+16], various emerging technologies have been studied and were shown to provide inherent security as IP protection schemes to hardware systems. Emerging nanotechnologies offer interesting transistor-level properties which can be explored for harnessing hardware security. Emerging technologies such as memristors [Raj+15] and spin-based devices [Pat+18a] have been explored with security-based features. Features like polymorphic gates and camouflaging in various emerging technologies [Kne20] can be extensively used in providing extra security to the IP at relatively less or no additional overhead.

Reconfigurable nanotechnologies discussed in this thesis, particularly, offer functional polymorphism inherently at the device level. Exploiting the possibility that a single device can be configured as a p-FET or n-FET at runtime, Bi et al. first suggested the concept of a key for RFET-based circuits that either activates the IC or makes the IC perform a specific function [Bi+14]. Similarly, some other works such as [Bi+16a, Che+16, RRK19] proposed exploiting emerging reconfigurable nanotechnologies using RFETs. However, most of these earlier works were ensemble works where various ideas were proposed using emerging technologies. SiNW RFETs [Mar+12] and tunnel FETs such as Graphene SymFets [Sed+14] are successful in providing an IC with the added feature of camouflaging when used with some specially designed structures. However, most of these works [Bi+14, Bi+16b, Che+16] lacked extensive evaluations for the proposed schemes with respect to the state-of-the-art defense or attack methods. This chapter focuses on laying experimental evaluations and providing empirical guarantees for security solutions with RFETs-based circuits.

7.4 Background

In this section, IP protection schemes towards hardware security are introduced. Security measures such as logic locking, hardware watermarking, and split manufacturing along with their threat models are discussed.

7.4.1 IP Protection Schemes

Designers worldwide see the need to incorporate security measures in their designs and verification; hence, adding security benefits is a cost. However, this distributed setup of fabrication, foundry, and testing allows several points where adversaries can attack the chip design or make illicit copies. With the growing proliferation of electronic circuits owing to the demand for autonomous vehicles and IoT applications, hardware security needs to be ensured at all levels of abstraction [Yas+19].

Various attacks such as the inclusion of hardware Trojans, IP piracy, IC overbuilding, reverse engineering, IC counterfeiting, and side-channel attacks are prevalent which can be detrimental for both the design houses and the end user. While the design houses face the issue of infringement of their intellectual properties (IP), the end user faces the ultimate risk of losing secret or private information. Researchers around the world have worked on various countermeasures like logic locking, IP watermarking, IC camouflaging, etc. to cope with such attacks [RKK14]. Within the context of RFETs, the present chapter focuses on introducing the concepts of logic locking and split manufacturing.

In this section, the background of logic locking and split manufacturing along with their related threat models is introduced. This sets up the foundation required to understand the security measures developed using RFETs.

Logic Locking

Logic locking has emerged as a viable and promising solution for protecting the design IP throughout the IC supply chain [Cha+20]. The IP is protected by the insertion of dedicated locks that are operated by a secret key. Therefore, a locked circuit has additional inputs, which are referred to as key inputs; these key inputs are driven by an on-chip tamper-proof memory. The additional logic gates inserted to realize these locks are known as key gates. Traditionally, these locks have been realized by adding XOR/XNOR gates, AND/OR gates, or Look-Up Tables (LUTs). A logic-locked IC functions correctly only when the correct key is applied; in the event of a wrong key being fed to the circuit, the IC becomes nonfunctional. After implementing a given locking scheme, the design house sends the chip to a foundry for fabrication, potentially *untrustworthy*. Once the chip has been fabricated and tested (but before deployment), the locked IC is activated by loading the secret key onto the chip's dedicated, tamper-proof memory by some *trustworthy* entity. For in-depth coverage of logic locking and associated attacks, interested readers are referred to [Cha+20].

Figure 7.1a shows an original (unprotected) circuit and Fig. 7.1b shows its locked version in traditional CMOS through three XOR/XNOR key gates. One of the inputs of each key gate is driven by a wire from the original design, while the other input, referred to as the key input, is driven by a key bit stored in a *tamper-proof* memory.

In general, a *threat model* quantifies the attackers' capabilities and available resources for launching attacks. The threat model for logic locking enumerated next bears consonance with the academic community's assumptions.

1. The design house, designers, and Computer Aided Design (CAD) tools are considered *trustworthy*, whereas the foundry, the test facility, and the end user(s) are all considered *untrustworthy*.
2. The attackers know the locking scheme implemented by the design house.
3. The attackers have access to the locked gate-level netlist (e.g., by reverse engineering) and can identify the key inputs and key gates but are oblivious to the secret key.
4. The secret key which is stored in a *tamper-proof* memory cannot be tampered with.
5. The attackers possess a functional chip that can be bought from the open market. Only "black-box" usage of the chip is permitted, i.e., an attacker can only

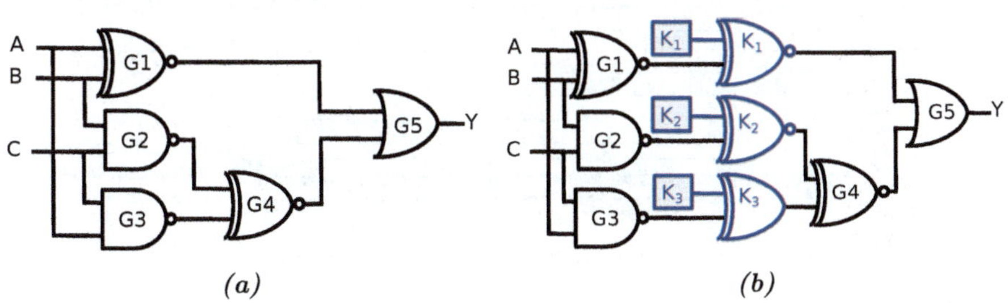

Fig. 7.1 Illustration of logic locking. (**a**) Original circuit. (**b**) Circuit locked with three XOR/XNOR key gates labeled as K_1, K_2, and K_3, respectively, for traditional CMOS technology. The key bits K_1, K_2, and K_3 are driven from an on-chip tamper-proof memory

evaluate input/output patterns. In recent terminologies, this kind of security attack is called as the *oracle-guided* attack scheme [Cha+20].

Split Manufacturing

Split manufacturing helps in the protection of the design IP from untrustworthy foundries during manufacturing time [Pat+18c, Pat+18b, Sen+19]. The split manufacturing premise dictates splitting up the IC manufacturing flow, typically into the Front-End-Of-Line (FEOL) and Back-End-Of-Line (BEOL). An attacker in the Front-End-Of-Line (FEOL) foundry views the layout as a "sea of unconnected gates" where some of the connections are complete and some of the connections are missing/incomplete. The notion of splitting the IC manufacturing flow into FEOL and Back-End-Of-Line (BEOL) is practical for multiple reasons: (i) outsourcing the FEOL is desired, since it necessitates advanced, high-end, and costly fabs, (ii) BEOL fabrication on top of the incomplete FEOL layout is significantly less complicated than FEOL fabrication, and (iii) the sole difference for the supply chain is the preparation and shipping of FEOL wafers to the BEOL facility. Figure 7.2 illustrates the idea of classical split manufacturing where the FEOL is outsourced to an advanced, offshore, *untrustworthy* foundry, while the BEOL is fabricated at a *trustworthy* foundry. For the in-depth coverage of split manufacturing and associated attacks, interested readers are referred to [KPS19].

The most common threat model adopted for split manufacturing is summarized as follows:

Fig. 7.2 Concept of classical split manufacturing, i.e., the separation of a physical layout into (FEOL) and back-end-of-line (BEOL). The different pitches across the metal layers facilitate split manufacturing. © 2018 IEEE. Reprinted, with permission, from [Pat+18c]

1. The design house, designers, CAD tools, and end user are *trustworthy*, while the FEOL foundry is considered *untrustworthy*. Split manufacturing dictates the existence of a BEOL foundry, with assembly and testing facilities, also considered as *trustworthy*.
2. The attackers cannot obtain a functional chip from the open market, as the end user is trusted. This scenario exists for military applications where the chips are deployed only for specific sensitive and mission-critical applications and are not available in open markets. Also, the chip has not been fabricated before and is unavailable for reverse engineering-based attacks.
3. The objective for an attacker (located in the FEOL foundry) is to infer the missing BEOL connections from the incomplete FEOL layout. Towards this end, the attacker is aware of the underlying protection schemes (if any) and has access to the EDA tools, libraries, and other information available to a trustworthy designer/design house.

RFET-Based Hardware Security

Runtime reconfiguration in RFETs enables the design of polymorphic logic gates, which, unlike regular CMOS logic gates, can perform more than one fixed logic function. Such polymorphic gates are promising for security schemes like camouflaging, logic locking, Physically-Unclonable Function (PUF), etc. The authors in [Bi+16b, Che+16] laid the foundation for hardware security using RFET-based circuits. Due to their inherent virtues of uniform physical layouts and post-manufacturing reconfigurability, such logic gates enable a security-enforcing designer to carefully obfuscate the circuitry from malicious foundries, test facilities, and/or end users. In turn, this hinders the theft of the chip's IP by adversaries. They also showed how a circuit layout composed with such RFET-based logic gates is difficult to reverse engineer. They demonstrated how a single tile layout could implement either a *NAND* or *XOR* using different pin configurations. However, these studies lacked both circuit-level simulations and thorough security evaluations against state-of-the-art attacks, e.g., the seminal SAT-based attack [SRM15]. Other hardware security schemes, such as watermarking using RFETs-based circuits, have also been proposed in [Rai+19a]. Similarly, RFET-based logic gates are less prone to delay-side-channel attacks as their CMOS counterpart [GG18, Sha+20] because of symmetrical electrical properties of an RFET in both n-type and p-type behaviors.

7.4.2 Preliminaries

The section describes the terminologies used in the chapter.

Hamming Distance

Hamming Distance (HD) is a metric used to compare two bitstreams. For two bitstreams of equal length, Hamming distance is the number of bit positions at which the two bitstreams are different. In terms of hardware security, it quantifies the average bit-level mismatch between two bitstreams obtained during attack or defense schemes. HD refers to the degree of functional mismatch; an HD value of 50% is considered the best-case scenario [Cha+20].

Output Error Rate

The Output Error Rate (OER) indicates the probability for any bit per output being different while applying an attack or defense scheme on a particular circuit (or netlist).

7.5 Security Promises

In this section, the security promises offered by RFETs due to their transistor-level reconfigurability are discussed. Detailed benchmark-level evaluations are presented in Sect. 7.7.1.

7.5.1 RFETs for Logic Locking (Transistor-Level Locking)

Figure 7.3 illustrates logic locking using RFETs. It should be noted that realizing logic locking in conventional CMOS necessitates the insertion of additional logic gates (e.g., XOR/XNOR, look-up tables [LUTs], etc.). In contrast, the RFETs-based circuits do not require the insertion of additional gates to realize locking. This is because RFETs have program gate (PG) terminals, which act as an in-built key, and hence, this notion of leveraging RFETs for logic locking can also be viewed as Transistor-Level Locking (TLL). In general, RFET-based logic locking offers lower area overheads than traditional logic-locking schemes since it does not require additional logic gates. Furthermore, since no additional logic gates are inserted, the original circuit's critical path continues to have the same number of stages as in the original circuit, thereby having no performance penalty.

With regards to area estimation, let us take an example of a circuit to be logic locked with k-bits. In the case of CMOS-based circuits, k logic gates (XOR/XNOR) are added for logic locking [Cha+20]. For CMOS circuits, each XOR consists of 10 transistors. Hence, for k-bit locking, the circuit will require $10 \times A_{CMOS} \times k$ number

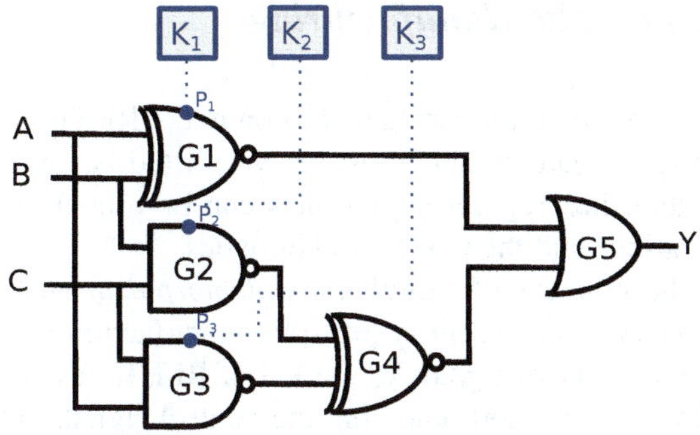

Fig. 7.3 Logic locking using RFETs, where the program gate (PG) acts as the key input. Realizing logic locking in conventional CMOS necessitates insertion of additional logic gates, whereas in case of RFETs, the program gate can enable logic locking without requiring insertion of additional logic gates. The PG signals are driven from an on-chip tamper-proof memory

of more transistors, where A_{CMOS} is the size of a CMOS transistor. For RFET-based circuits, since they allow inherent reconfiguration, for k-bit logic locking, it will require k inverters to have P (and P') as the extra input for each reconfigurable logic gate [Rai+19b]. Each inverter consists of 2 transistors. Hence, the area overhead for the same locking scheme is $2 \times A_{RFET} \times k$, where A_{RFET} is the size of an individual RFET transistor. Between the technologies, the overall area overhead is calculated as:

$$Overhead = (n_{CMOS} \times A_{CMOS} - n_{RFET} \times A_{RFET}) \\ + (10 \times A_{CMOS} \times k - 2 \times A_{RFET} \times k) \quad (7.1)$$

Here, n_{CMOS} and n_{RFET} are the number of transistors in CMOS and RFET-based circuits, respectively. The first term in Eq. 7.1 is the difference in the area of the same circuit in two technologies, while the second term is the actual overhead for ensuring k-bit locking. One can notice that for k-bit locking, the RFET overhead is less than that of CMOS by a factor of 5 (if $A_{CMOS} = A_{RFET}$). Hence, the overall area overhead depends upon two main factors—(i) the number of transistors in a circuit and (ii) the size of an individual transistor. Due to higher functional expression, the number of transistors in case of RFET-based circuits is much less as compared to CMOS-based circuits [Rai+19b, Zha+14], i.e., $n_{RFET} < n_{CMOS}$. Since RFET models are still evolving, with better RFET models, A_{RFET} will get closer to A_{CMOS}. For instance, for an early evaluation model of RFET, it was demonstrated in [RRK18] that for the same circuit, the actual area for RFET-based circuits is just 17% more than that of CMOS. Bringing all these factors in consideration and when using the same technology node, an RFET-based circuit has a lower area as compared to a CMOS circuit for the same k-bit logic locking.

7.5.2 RFETs for Split Manufacturing

The applicability of split manufacturing in the context of RFETs is shown in Fig. 7.4. Here, only the program gate signals must be wire lifted beyond the split layer. In contrast, all the other, i.e., regular signal, nets can be routed freely following the designer's specifications and the CAD tool heuristics.

In this work, the concept of *functional polymorphism* enabled by RFETs is leveraged for enhancing the security of split manufacturing. As indicated, the application of various control signals to the PG of RFETs results in different logic functionalities. However, without knowing the control signal, one cannot infer the logic gate's actual functionality. Moreover, the concept of logic locking is utilized towards securing split manufacturing.

Both concepts taken together, polymorphism and locking, allow us to assign individual key bits directly for any logic gate of choice, but it is also required to lift the related wiring associated with the program gates above the split layer. This way, the actual functionality is obfuscated of these gates from the foundry-based adversaries. The essence of this approach is similar to a recently published work [Sen+19] where the authors inserted key- gates to lock the FEOL layout and then controlled the routing of the related key nets through the BEOL. However,

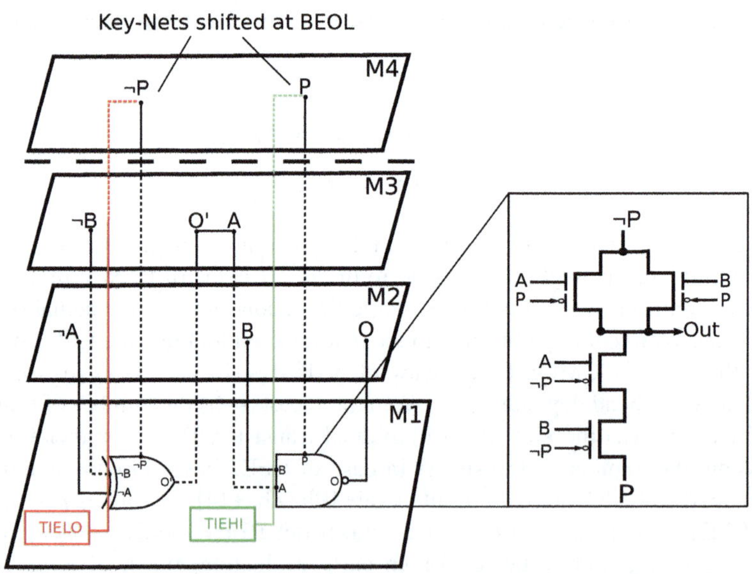

Fig. 7.4 Split manufacturing for RFETs; here, only the program gate (PG) signals are lifted beyond the split layer (in this example, M3) to BEOL (M4). The inset shows the internal transistor-level schematic of the underlying logic gate. These lifted PG signals will be driven by constant 0/1 signals (generated by TIELO/TIEHI cells placed in the FEOL), routed in the BEOL. The absence of placement- and routing-related hints for the key nets makes the key indecipherable for an FEOL-centric attacker

there is one crucial difference between the approach [Sen+19] and the proposed approach is that the approach in [Sen+19] relies on the insertion of additional logic gates to achieve the required security guarantees, but our scheme benefits from the inherent polymorphism of the RFETs, avoiding any additional modifications while still imposing strong security.

7.6 Security Vulnerabilities

While the previous section focused on the security promises offered by RFETs, in the present section, the security vulnerabilities are demonstrated for RFET-based circuits. Detailed circuit-level simulations are conducted in this section followed by further analytical studies in Sect. 7.7.2. It becomes interesting to note that while the same feature of *functional polymorphism* contributes to making the circuit secure, it can be exploited by an opportune attacker for circuit degradation techniques.

7.6.1 Realization of Short-Circuit and Open-Circuit Scenarios in an RFET-Based Inverter

Conventionally, for CMOS-based logic gates, the p-channel Metal Oxide Semiconductor (PMOS) and n-channel Metal Oxide Semiconductor (NMOS) transistors realize their specific functionality in pull-up and pull-down networks, respectively. CMOS circuits require a separate pull-up and pull-down network for keeping the output either at logic high (1) or low (0). When the pull-up network is switched on (off), the output is pulled up (pulled down) to logic 1 (logic 0). Both networks are simultaneously on and current flows through the transistors only during the switching of the output from logic 0 to logic 1. However, in RFET-based logic gates, the boundary between PUNs and PDNs is somewhat blurred. That is, the same pull-up (pull-down) network can also work as a pull-down (pull-up) network by merely changing the configuration of all the related transistors. The drive strength for RFETs in both n-FET or p-FET configuration is identical [Tro+15], which makes this switching between pull-up and pull-down possible, to begin with. This specific property is the root cause of the security vulnerabilities discussed here.

More specifically, the potential exploit in RFET-based circuits arises from the fact that individual RFETs of a circuit can be individually programmed maliciously. RFETs can be used as p-type or n-type transistors, irrespective of their presence in the pull-up or pull-down network, as this behavior merely depends upon the potential at the PG of individual RFETs, as shown in Table 2.1. In essence, given this reconfigurability of individual RFETs, one can either switch "on" certain transistors in the pull-up or the pull-down network to induce a *short-circuit* path from V_{dd} to

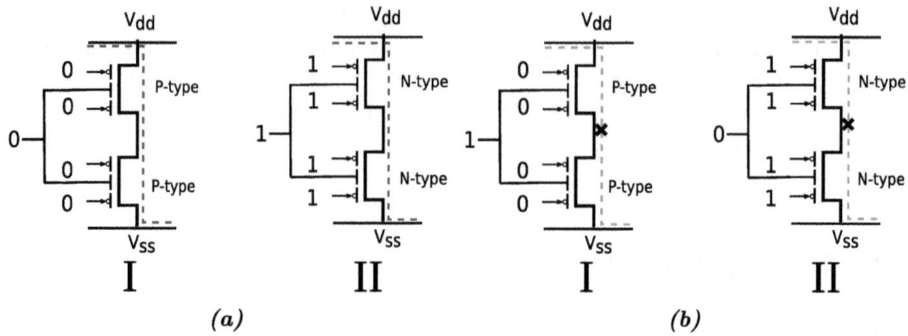

Fig. 7.5 Exemplifying security vulnerabilities in an RFET-based inverter: (**a**) two short-circuit configurations and (**b**) two open-circuit configurations

V_{ss} or switch "off" certain transistors to induce an *open-circuit* between V_{dd} and V_{ss}, respectively.

This is explained by an example in Fig. 7.5. The figure shows a normal inverter, where particular "misconfigurations" lead to unfavorable conditions. Figure 7.5a shows an inverter comprising of two RFETs where both the transistors are configured as either p-type (I) or n-type (II). This is possible when both the gate terminals, PG and CG, are assigned the same potential. In these cases, both the transistors are switched on, leading to a conducting path between the V_{dd} and the V_{ss} nodes. We refer to this configuration as the *short-circuit configuration.* The reverse scenario, shown in Fig. 7.5b, occurs when the potential at the PG and CG is inverse of each other. In these cases, both the transistors are switched off, leading to an *open-circuit configuration.* The red dashed lines in Fig. 7.5a signify a short-circuit path, while grey dashed lines in Fig. 7.5b signify an open-circuit path.

To understand the behavior in the two conditions mentioned above, we study an RFET-based inverter (shown in Fig. 7.6a) and compare its normal operation to the case when it is misconfigured. We explore scenarios when a potential bias is swept across the gate terminal of an inverter. The simulations are performed using *Cadence Virtuoso* with a simple table-based RFET model used in [Gor+19].

Figure 7.6b shows the amount of current drawn from the inverter in which both the transistors are *on* and are either fixed as n-type or p-type (indicated by blue and red curves, respectively). The blue curve shows the condition in which the control gate input to the inverter is fixed at logic 1, while the program gate input of the lower transistor, Q, is changed from logic 0 to logic 1, signifying reconfiguration from p-type to n-type. Similarly, the red curve exhibits the case when the control gate input to the inverter is fixed at logic 0. Simultaneously, the program gate input P is varied from logic 1 to logic 0, signifying reconfiguration from n-type to p-type. Both the cases can create a short-circuit or open-circuit path between V_{dd} and V_{ss} depending upon the input at A.

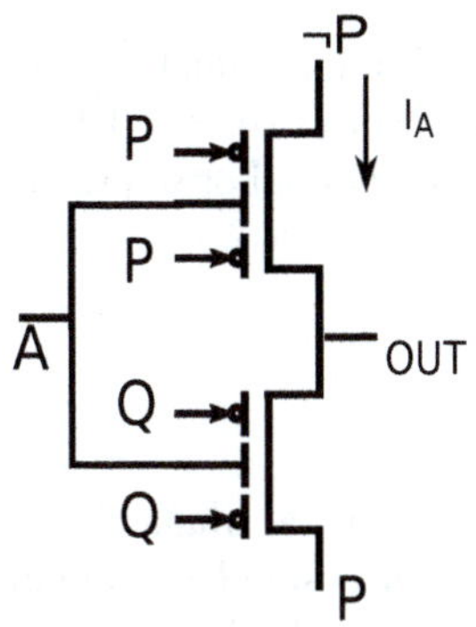

(a)

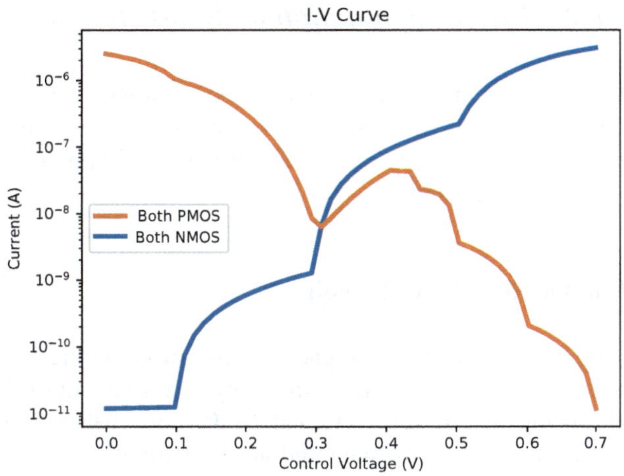

(b)

Fig. 7.6 (**a**) Short-circuit operation possible in case of an inverter by switching on both the transistor in either p-type or n-type at runtime. (**b**) Simulation results showing current reaching 10^{-7}A range for inverter in two configurations

1. *Large short-circuit currents*: One can see from Fig. 7.6b that the amount of current flowing through the inverter during the short-circuit condition is of the order of 10^{-6}A in both the cases, which is almost 10^5 times higher than the static leakage current of 10^{-11}A under normal operating conditions for RFETs [Gor+19]. The unique property of RFETs is that any logic gate can be similarly reconfigured at runtime, which gives rise to the possibility of creating multiple short circuits, resulting in a large amount of current being drawn from the power source. Other unwanted effects include high-power dissipation and excessive localized heating near the affected portion of the circuit. Such a scenario could lead to accelerated aging of the circuit and critical reliability issues like the thermal shutdown of the system [Das14]. The increase in power dissipation will eventually lead to a reduction in the battery lifetime, especially for portable devices like laptops, mobile phones, etc.
2. *Fault in logic values*: In the cases shown in Fig. 7.5a and b, the output of the inverter will be at an indeterminate voltage level. If this misconfigured inverter drives a gate, it can have faulty inputs and evaluate wrong logic outputs. This phenomenon can continue further in a combinational circuit leading to a chain reaction where preceding nodes can lead to indeterminate voltages at subsequent nodes. Eventually, this would lead to incorrect functioning of the overall circuit.

7.6.2 *Circuit Evaluation on Sub-Circuits*

We notice the current and voltage repercussions in the case of an individual inverter. This section discusses how such a misconfiguration in an inverter (or any RFET-based logic gate) can disrupt the normal functioning of combinational or sequential sub-circuit elements.

Combinational Sub-Circuits

Here we evaluate the impact of the misconfigured inverter in a small circuit, as shown in Fig. 7.7. For simplicity, we study the effect on a small sub-circuit consisting of NAND, INV, and NOR. Such behavior is triggered when the gate inputs are in a steady state and not switching.[1] The NAND gate on the left drives the misconfigured inverter, which in turn drives the NOR gate. The program input Q of the inverter (as discussed in the case study above) is fixed at logic '1' (for n-type configuration) for it to work undetected as a regular inverter. When the inverter is misconfigured, the current and voltage values for all the possible configurations and inputs are listed in Table 7.1. It can be observed from the table that current values reach to orders as high as 10^{-6}A, as opposed to 10^{-12}A under regular, static operation. This million times more current can potentially damage the power source and cause thermal issues and affect the logical output of the overall circuit. The implication of the security vulnerability in the case of RFETs is somewhat similar

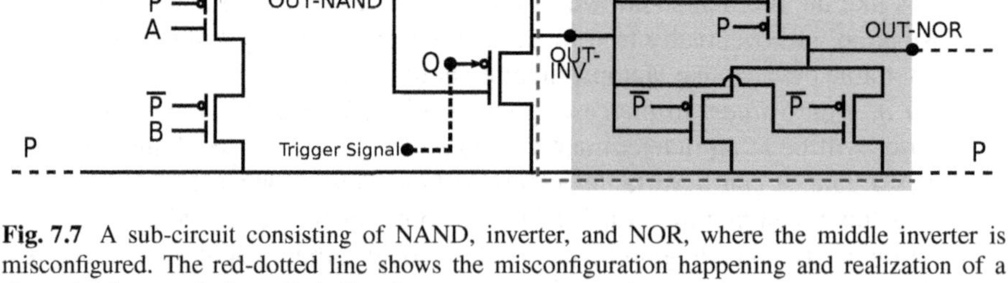

Fig. 7.7 A sub-circuit consisting of NAND, inverter, and NOR, where the middle inverter is misconfigured. The red-dotted line shows the misconfiguration happening and realization of a short-circuit scenario through the inverter

[1] This is assumed considering the fact that once the scenario of *short circuit* or *open circuit* is activated, irrespective of the previous state, the logic gate will be in one of the conditions as mentioned in Table 7.1.

to the issue of *latchup*, which was a serious problem with CMOS integration in the early days of VLSI fabrication [Das14]. *Latchup* was caused due to faulty transistors in either pull-up or pull-down network, which caused current discharge from V_{dd} to V_{ss}.

The activation of such a scenario also leads to the node *OUT-INV* in Fig. 7.7 to be at an undefined state, which affects the input of the subsequent NOR gate. It implies that the subsequent gates in the combinational path have the likelihood of going into an indeterminate state (NOR's output can affect subsequent logic gates). This can cause unpredictable bit-flips along the way, affecting the functional correctness of the circuit. Table 7.1 shows the voltage levels at various nodes of the circuit in Fig. 7.7 for all possibilities of inputs. Especially, it can be remarked from the table that the *OUT-NOR* node voltages for *both p-type* (*both n-type*) configurations are held at logic high (low) when compared to the one under normal operation. This further strengthens the point that such a malicious modification can interfere with the functional correctness of the circuit.

It can be further observed that such a scenario is possible in other combinational chains and is independent of the logic gate used as shown in Table 7.1. The inverter is a special case because here, we need only one input of an individual transistor (Q, as shown in Fig. 7.7) to be wrongly configured. Larger gates like *XOR*, etc., can also be disrupted as it depends upon the number of transistors which are wrongly configured to enable the realization of such *short-circuit* or *open-circuit* paths.

Sequential sub-Circuit

We carry evaluations over sequential components as well since most of the real circuits are sequential circuits. For this analysis, we have considered the RFET-based *true single-phase clock (TSPC) D-flip-flop* as proposed in [Zha+14], which is based on the original CMOS-based TSPC DFF [JKS87]. It is shown in Fig. 7.8.

Unlike the original design [JKS87], the proposed design in [Zha+14] contains only a single transistor in both pull-up and pull-down networks.[2] The TSPC-DFF contains four stages of different inverter designs. While the first- and the third-stage inverters are clock-enabled inverters (low- and high-edge enabled, respectively), the second and the fourth inverter stages are dynamic (CLK is connected to the CG of the transistor) and static inverters, respectively.

To observe the effects of RFET-based vulnerabilities, we evaluate by introducing misconfigurations at the gate terminals of RFETs of the various stages of the TSPC-DFF. In Fig. 7.8, the transistors' configuration (shown in red color) in the second and third stage has been changed subsequently. When applied in isolation (the configurations are changed either in the second stage or the third stage), both conditions lead to abnormal outputs being observed at the output Q along with

[2] This is because RFETs can have multi-independent terminals on a single channel as explained in Chap. 2.

Table 7.1 Simulation results showing current drawn and voltage values for different cases of inputs and configurations

	Overall current drawn through INV			At node OUT-INV			At final output		
	Normal operation	Both n-type	Both p-type	Normal operation	Both n-type	Both p-type	Normal operation	Both n-type	Both p-type
INV-INV-INV									
Output-1 = 0 V	389 pA	3.05 uA	8.19 pA	0 V	31.13 mV	686 mV	0 V	700 mV	0 V
Output-1 = 700 mV	83 pA	251 pA	2.5 uA	700 mV	106.7 mV	609 mV	700 mV	700 mV	0 V
XOR-INV-NAND									
OUT-XOR = 0 V	414 pA	251.8 pA	2.5 uA	700 mV	31.06 mV	610 mV	0 V	700 mV	0 V
OUT-XOR = 700 mV	11.58 pA	3.0 uA	7.17 pA	0 V	100 mV	688 mV	700 mV	700 mV	0 V
NAND-INV-NOR									
OUT-NAND = 0 V	414 pA	251 pA	2.5 uA	700 mV	31 mV	609 mV	0 V	700 mV	0 V
OUT-NAND = 700 mV	11.8 pA	3.055 uA	29.01 pA	0 V	106 mV	688 mV	700 mV	700 mV	0 V

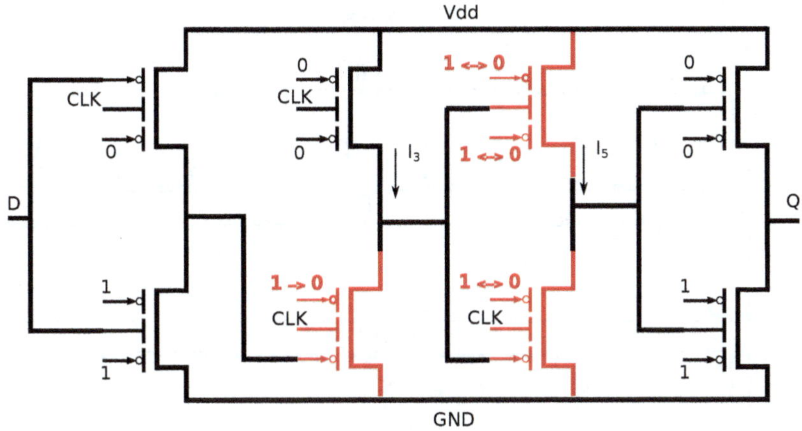

Fig. 7.8 True single-phase clock (TSPC)-DFF based on RFETs as proposed in [Zha+14]. Just by changing the polarity of the lower transistor at the second stage (precharge stage when CLK = 0 and evaluation phase when CLK = 1)

a large amount of current flowing through the second and third stages. We study the effects when the transistors of the second and the third stage are independently configured in a wrong manner.

Effect on Voltage Figure 7.9 shows voltages at the output Q in various conditions. In Fig. 7.9, the normal Q represents the condition when the TSPC-DFF functions normally as the output Q follows D at the next positive edge of clk. The next figure represented by *2-ab* Q (ab = abnormal case) shows when the transistor in the second stage is misconfigured, here as p-FET. The output Q remains at logic high. Similarly, when both transistors at the third stage are either configured as p-FET or n-FET, the output Q is represented in Fig. 7.9 as *3-ab-P* Q or *3-ab-N* Q.

This corrupts the output Q, which is seen to be at logic 0 since the pull-up network never conducts in the fourth stage. Similarly, when both the transistors in the third stage are configured as NMOS, we observe that output Q follows the clock as a stable pull-up ceases to exist for the fourth stage. We can see that due to the incorrect configurations, the TSPC-DFF fails to give the desired output.

Effect on Current Figure 7.10 shows the current through various paths of TSPC-DFF. A corresponding correlation with the voltage can be established for the currents I_3 and I_5. While the currents I_3 and I_5 in normal scenario show spikes for the second rising edge of the clock, this is highly disrupted when the second- or third-stage transistors are misconfigured. When the transistor in the second stage is misconfigured, there are several spikes of currents (both I_3 and I_5), which lead to useless dynamic power consumption and potential reliability concerns. Hence, such RFET vulnerabilities can be exploited for both combinational and sequential components of the circuit.

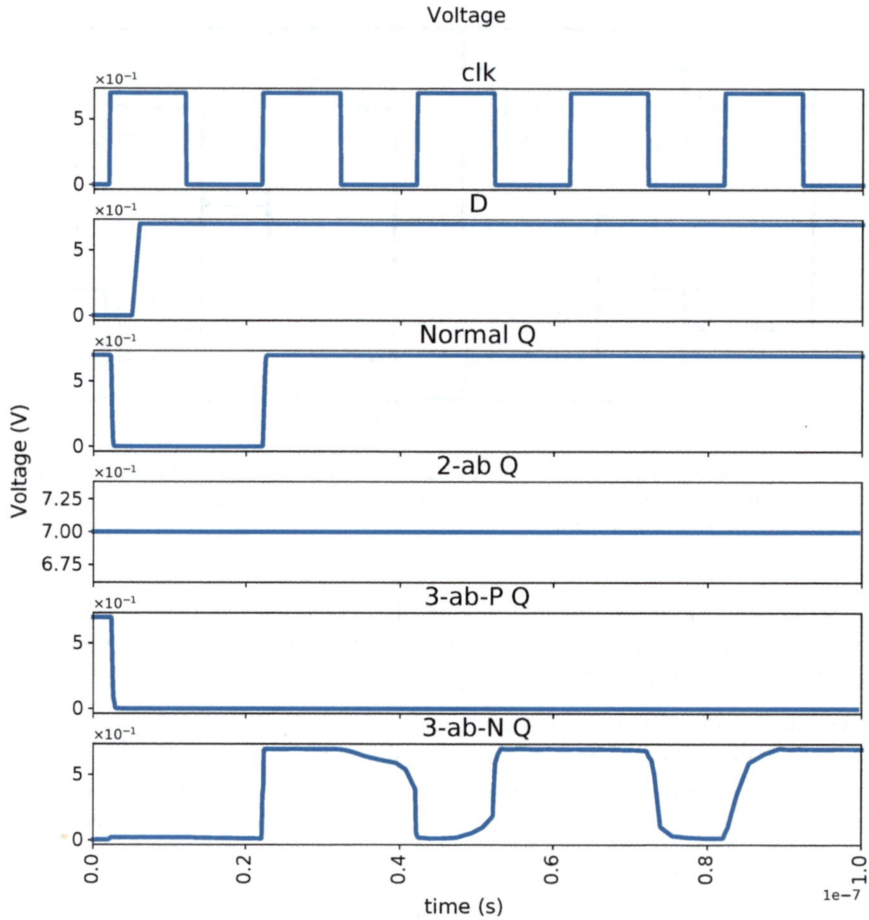

Fig. 7.9 Voltage values in normal and both short-circuit and open-circuit conditions for TSPC-DFF. 2-ab means when the transistor in the second stage is configured incorrectly. 3-ab-N and 3-ab-P imply that both the transistors in the third stage are either configured as NMOS and PMOS, respectively

7.6.3 Reliability Concerns: A Consequence of Short-Circuit Scenario

We have seen that security vulnerabilities using RFETs can have a detrimental impact on both the voltage at the output stage and the current in one or multiple paths in the circuit. With devices based on emerging nanotechnologies, where individual transistors can be configured either as a p-type or an n-type behavior, current-based implications can be detrimental, especially for reliability. We have seen that the current surge of the order of 10^5–10^6 is possible with RFET-based circuits. Such

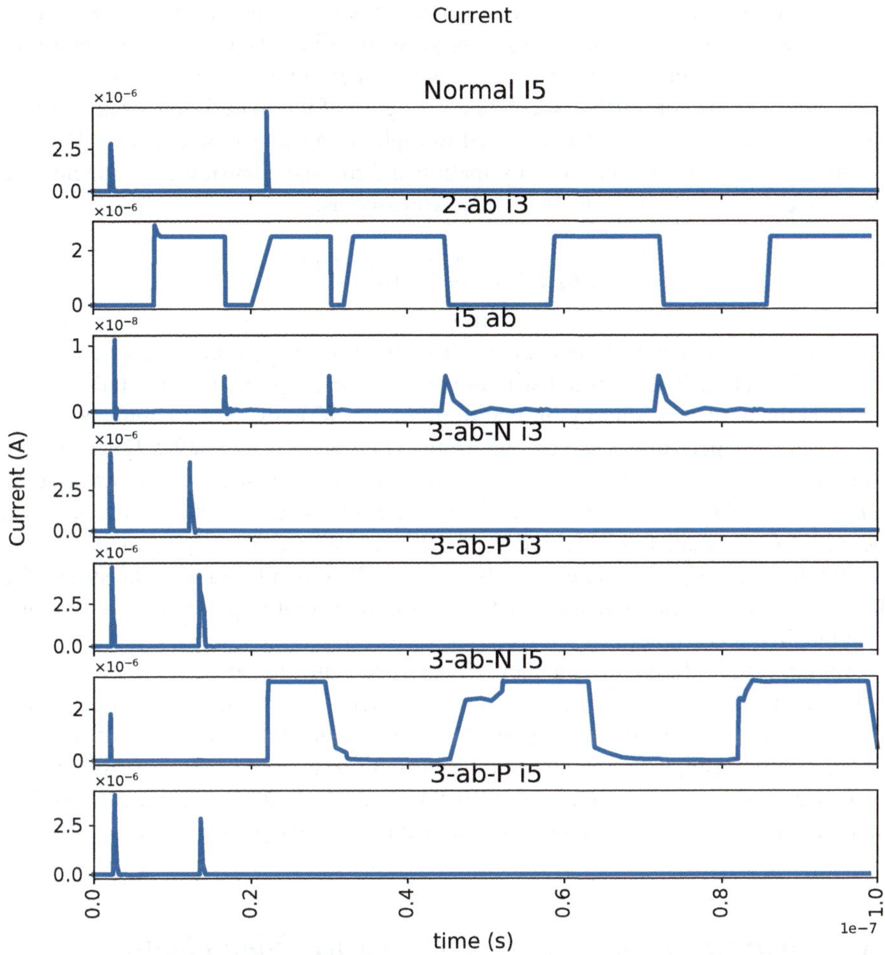

Fig. 7.10 Current values in normal and both short-circuit and open-circuit conditions for TSPC-DFF. 2-ab means when the transistor in the second stage is configured incorrectly. 3-ab-N and 3-ab-P imply that both the transistors in the third stage are either configured as NMOS and PMOS, respectively

current surges in one or more of the circuit paths can induce reliability issues that can manifest themselves leading to a reduced quality of service (*QoS*). An increased current through the circuit paths for an extended time can further lead to an increase in temperature. Higher current leads to a similar mechanical stress which can cause a material degradation of the transistors. Since RFETs are similar to CMOS in terms of geometry and material, they also have dielectric separation (HfO_2) between metal contacts [Mik+17]. Such current-related issues directly impact the dielectric and other metal-semiconductor contacts and, hence, are also detrimental in case of RFET-based circuits.

Most of the reliability issues are dependent upon current and temperature. Due to increased temperature and current, prominent reliability issues like electromigration, interconnect, and self-heating can surface, derailing the circuit's normal functionality. This can further accelerate the aging of the underlying circuit. Aging happens in circuits due to stress related to high voltages or temperatures [Das14]. These can cause unwanted power dissipation and can also contribute to reducing the *mean time to failure* $(MTTF)$, which is expressed as:

$$MTTF = \frac{A_{EM}}{J^n} \cdot 10^{\left(\frac{E_{aEM}}{KT}\right)} \tag{7.2}$$

where A_{EM} is a material-dependent constant, J is the current density, n is an empirically determined constant with a typical value of 2 for stress-related failures, E_{aEM} is the activation energy of electromigration, K is the Boltzmann's constant, and T is the temperature. It can be seen from expression 7.2 that $MTTF$ is inversely proportional to the square of the current density (J^2). Hence, even if we assume temperature to be constant (which is the worst-case scenario), with a current surge, the $MTTF$ is accelerated by order of 10^{10} to 10^{12}. Generally, as a basic rule of thumb, $MTTF$ is 10 years. In the case of RFET-induced current surge, the $MTTF$ reduces from 10 years to 0.3 s. This implies that in just 0.3 s, the circuit's functionality is corrupted.

An important thing to note here is that the current surge's effect is more detrimental and pervasive to the circuit compared to the voltage effects. Voltage effects might be masked by other parts of the circuit and may not present itself at the circuit's output. But, in the scenario of the current surge, once triggered, this can ruin the underlying circuit even if the circuit functions according to its specifications. The silent nature of this effect is more adverse from a security point of view.

7.6.4 Implication of the Proposed Security Vulnerability

Normally, a current flowing higher than the rated current (the on-state current in this regard) is capable of damaging the transistors. Moreover, it is possible that by attacking a small portion of the actual chip, we may be able to ruin the functionality and appearance of the chip as well. This technology may find intensive application in fields where the security of IP is of utmost importance. Military is one such field. One of the major implications of the proposed security vulnerability is the realization of *Hardware Trojans*. Given that the pull-up/pull-down network in an RFET-based logic gate could be falsely configured, any gate could become a Trojan in the field. Hardware Trojans present security concern in the outsourced IC supply chain. Trojans are malicious modifications inserted by adversaries present either in the design house or at the foundry which can cause the host circuit to (i) deviate from their specified functionality, (ii) leak sensitive information, and/or (iii) become unreliable or fail at some point in time [Xia+16]. The proposed security vulnerability can thus be exploited to realize reliability Trojans which are activated either by (i) aging effects such as electromigration or (ii) internal or external side-channel triggers. Such benign or reliability hardware Trojans have been explored before in the literature and can compromise the reliability of all or selected chips in an SOC [Shi+10, Xia+16].

An integral part of such Trojans is to use triggering mechanisms that remain inactive during the testing phase, triggered by some external or internal trigger conditions at any given point in time. Within the context of RFET-based circuit, it is important to note that Trojan realization in this case is different from traditional Trojans, as it can evade detection during the testing phase. For example, *Iddq* testing,[3] which can detect electrical faults, will not be able to identify these scenarios (*short circuit/open circuit*) because, in the normal case, when RFETs are not misconfigured, there will not be any such electrical faults. The main contribution of this work, however, is the study of security vulnerability in RFET-based circuits, not the advancement of Trojan insertion or their defense mechanisms.

Another potential use of such security vulnerability is in the development of a *kill-switch*, which is any manipulation of the chip's software or hardware that causes the chip to shut down the intended functionality, for example, to shut down an F-35's missile-launching electronics [Ade08]. The above scenarios of short circuit and open circuit represent an interesting opportunity for a security-enforcing designer. The ability to configure the RFETs at runtime to seriously disrupt the normal functioning can be highly favorable for safety-critical applications such as for military use.

In order to tackle such security vulnerability, it is important to use RFETs' functional polymorphism in a controlled way. Static connections to the program gate of individual transistors within a single logic gate should be avoided. Dynamic connections using inverter(s) within the gate boundary to enable both P and P' signals can be one of the ways to get away with security vulnerability. The use of inverter(s) within the gate boundary implies that P's additional input is fed as an extra input to the enlarged standard cell. The inverter then converts the input P to deliver two signals—P and P'—that drive the RFETs accordingly to ensure that the respective pull-up and pull-down networks are intact. This ensures that no external signal can come into the gate to disrupt the complementary pull-up and pull-down networks. However, this approach does have the drawback of higher area overheads due to extra inverters necessary within the gate boundary.

Other measures include devising *guard rings* [Dom18] or *voltage controllers* [Svi+10] to localize the vulnerability effect to a certain part of the circuit. Guard rings simply help to localize the current drawn from V_{dd} and can be placed for few RFETs together. They also help to provide a low-resistance path for the current to flow without harming other parts of the circuit. Similarly, voltage controllers can

[3] *Iddq* testing measures the supply current of a chip (or a given module) in the quiescent state (i.e., when the circuit is not switching and inputs are held at static/constant values) to detect manufacturing and/or electrical defects.

be activated to cut off certain sections of the circuits from the rest of the circuits in case of excessive current drawn from the voltage source. However, these methods are preventive orthogonal measures that help only after the attack has been in effect so as to localize and prevent the damage to other parts of the circuits. They also have additional area and power overheads. More importantly, they have been designed primarily for signal correctness in conventional CMOS circuitry and were generally designed separately for p- and n-type transistors. However, in case of RFETs, where such a boundary is blurred, devising such mechanisms require additional investigation.

7.7 Analytical Evaluation

7.7.1 Investigating the Security Promises

In this section, the findings of the evaluation security promises using RFETs are discussed. We explore the security guarantees by leveraging RFETs in logic locking and split manufacturing.

Setup for Security Evaluation For the scenario of malicious end users, the locking approach is evaluated against powerful *exact* SAT-based attacks [SRM15] and *approximate* SAT-based attacks [Sha+17]. Different transistor-level configurations for RFETs are assumed, which can work as NAND/NOR, AND/OR, and XOR/XNOR. RFET-based locking is implemented as individual 2-to-1 MUXes as outlined in [Cha+20]. Since both the SAT-based attacks (exact and approximate) require the netlists to be in *BENCH* format, custom *Python* scripts are employed to implement the locking approach. The SAT-based attacks [SRM15, Sha+17] are carried out on a server with five compute nodes; each node has two 14-core Intel Broadwell processors, running at 2.4 GHz with 128-GB RAM. The time-out for the attacks ("t-o") is set to 48 h. Random logic locking is implemented for experiments; ten different sets for each benchmark is generated, ranging from 10% to 50% locking, in steps of 10%.

Metrics for Security Evaluation The attacks' average runtime is attributed as an empirical, yet essential indicator for a design's resilience. Two well-known metrics are utilized—HD and OER to evaluate the quality of netlists inferred by successful attack runs. HD and OERare computed using *Synopsys VCS* and functional correctness of the key (dumped by the exact SAT-based attack [SRM15]) is ascertained by *Synopsys Formality* and *Cadence LEC*.

Results for Logic Locking Table 7.2 illustrates the average runtime (in seconds) required for SAT-based attacks [SRM15] to decipher the locking key for randomly

Table 7.2 Average runtime (in seconds) for SAT-based attacks [SRM15] for different randomly locked configurations on selected *ITC-99* benchmarks. Time-out (t-o) is 48 h. For each benchmark ten trials are used

Benchmark	10% locking	30% locking	50% locking
b14_C	18.43	6,183.84	t-o
b15_C	21.47	5,398.43	t-o
b17_C	312.32	21,324.76	t-o
b18_C	1,543.73	156,378.23	t-o
b19_C	4,678.79	t-o	t-o
b22_C	143.21	9,762.74	t-o

locked configurations on selected *ITC-99* benchmarks.[4] It is observed that as the number of locked gates is increased (50% locking), the SAT-based attack cannot decipher the locking key within 48 h. The resilience of the large-scale locking approach is also examined against approximate key recovery attacks like *AppSAT* [Sha+17]. More specifically, regarding resilience against *AppSAT*, the attack on the same locked benchmarks is executed. While *AppSAT* ran into time-out after 48 h, it is programmed to provide its latest, best-as-possible inference as an approximate key before terminating. These keys are applied to (approximately) unlock the netlists and then the HD is calculated between this recovered and the original netlist. This provides the quantified insight into the recovered key's fidelity. The experiments are performed only for those cases where the exact SAT-based attack [SRM15] runs into time-out, that is, 50% locking for all *ITC-99* benchmarks. As expected, it is observed that the larger the locking scale, the recovered key becomes less useful, in the sense that the HD values approach 50% closely, where the mismatch in functional behavior is the most difficult to recover.

Security Promises Using RFETs for Split Manufacturing Here, we showcase the security promises where the unique properties of *functional reconfiguration* and *functional polymorphism* can be leveraged for protecting design IP against untrusted foundries using the concept of split manufacturing. As elucidated previously, a designer can protect the entire netlist by choosing RFETs that offer "fixed functionality" versus RFETs which provide "variable functionality" based on the control signals provided on the program gates (PG) [RRK18]. The conceptual difference with regards to logic locking is that, in the case of split manufacturing, the secret key shall be implemented via connections only in the BEOL with the help of *TIE* cells (as opposed to a tamper-proof memory). Towards that end, only the program gates of selected RFETs have to be wirelifted above the desired split layer (see Fig. 7.4) and driven with constant '0'/'1' signals, which are routed through the BEOL.

[4] It is to be noted that *ITC-99* benchmark suite is more representative of the security community particularly in works related to logic locking compared to let us say EPFL benchmarks [AGD15].

Table 7.3 Hamming distance (HD) and output error rate (OER) in percentage on selected *ITC-99* benchmarks for 1 million test patterns as a function of lifting of program gate signals

	10% lifting		30% lifting		50% lifting	
Benchmark	HD	OER	HD	OER	HD	OER
b14_C	16	100	34	100	47	100
b15_C	22	100	38	100	49	100
b17_C	19	100	34	100	43	100
b18_C	11	100	24	100	41	100
b19_C	12	100	23	100	39	100
b22_C	21	100	34	100	47	100
Average	17	100	31	100	44	100

Why Proximity Attacks Will Not Be Successful? The success of conventional *proximity attacks* is correlated with the type of FEOL-level information, which can be harnessed to infer the missing BEOL connections. More specifically, the state-of-the-art network-flow attack aims to reconstruct the missing routing of a FEOL design leveraging the following information like (i) physical proximity between connected cells; (ii) routing patterns of nets, more specifically, the direction of dangling nets in the FEOL; (iii) constraints of load capacitance for drivers; (iv) the non-formation of combinational loops; and (v) timing constraints.

None of the above information apply to the TIE cells (which supply a fixed '0' and '1', respectively to the PG) in the construction of the experiments. This is because physical proximity between the TIE cells and the corresponding PG signals of the RFETs can be eliminated by randomizing the TIE cells' placement. The randomization of TIE cells' placement does not impact the placement and/or routing of the underlying design. The information which emanate from the routing patterns of the FEOL nets can be further obscured by lifting the whole metal segment to the BEOL; note that the usage of stacked vias can be leveraged towards this end. As wirelift is applied only with a fixed number of PG signals, it would ensure minimal disturbance to the routing of the other regular nets in the design. Once the percentage of wires which are lifted across higher metal layers is increased, there would be an increase in congestion (due to scarcity of routing resources), which can be mitigated by increasing the die outlines. The information of load capacitance constraints does not apply to TIE cells as they are a source for constant signals like '0' and '1'. Since any other logic does not drive TIE cells, the information of non-formation of combinational loops is also taken care of by construction. Finally, the information for timing constraints does not apply to TIE cells (and nets) as they provide a fixed/static path to the PG signals.

Results for Split Manufacturing In general *proximity attacks* [Wan+18] are executed to ascertain the strength of any defense pertaining to split manufacturing [Pat+18c, Pat+18b, Sen+19]. However, the attack binary released in [Wan+18] cannot be ported readily to RFETs. Hence, to showcase the efficacy of RFETs for split manufacturing, a simple, yet effective experiment is conducted as follows. An assumption is taken that all the regular nets have been correctly inferred by an attacker (using some proximity attack of choice) and only the program gate (PG) signals remain to be deciphered. Since the threat model of split manufacturing dictates the nonavailability of a working chip, SAT-based attacks like [SRM15] are not directly applicable. At most, an attacker can apply "random guessing," and this is imitated by using 1 million test patterns on the netlists. Table 7.3 denotes the HD and OER obtained after an attack executed by attackers present in an untrusted foundry on selected *ITC-99* benchmarks as a function of the percentage of lifted program gate (PG) signals. As can be noted from Table 7.3, the HD value increases as program gate (PG) signals are lifted to higher layers. It can also be observed that OER approaches the ideal value of 100% even when 10% wires are lifted to higher metal layers, thereby showcasing the approach's efficacy.

7.7.2 Investigating the Security Vulnerabilities

A benchmark-based evaluation of the impact of the voltage effects of the security vulnerability on the functional correctness is carried out. Next, the experimentation process and the obtained results are presented.

Experimental Setup As it can be seen that with the creation of a *short-circuit* or an *open-circuit* path (see Fig. 7.5a, b, respectively), various nodes in a circuit can have indeterministic and random voltage. These nodes are termed as "misconfigured nodes." To model such indeterministic voltages in the experiments, *stuck-at-1* and *stuck-at-0* faults are fixed randomly at these misconfigured nodes in the gate-level netlist for *MCNC*, *EPFL* [AGD15], and *ITC-99* benchmark circuits. We use the fault analysis tool *HOPE* [LH96] to analyze these misconfigured nodes of a circuit.

Each benchmark is tested with 10,000 input patterns and the number of test patterns for which the outputs change is noted. To carry out analysis, 50 iterations have been carried out on randomly selected nodes (logic gates) in the netlist for *MCNC*, *EPFL*, and *ITC-99* benchmarks. The three benchmark suites are chosen as they represent the different sizes of electronic circuits. Similarly, the type of fault inserted at each node is also chosen arbitrarily.

Figure 7.11 shows OER and HD for 10,000 input test patterns for *MCNC*, *EPFL*, and *ITC-99* benchmarks. The results presented represent an average of the respective metrics over all 50 iterations. A more in-depth analysis of the benchmark-level evaluation is provided below.

Results The fault created as a result of misconfiguring various nodes in the circuit is said to create the maximum impact on the fidelity of the circuit's output response when exactly half of the output ports are affected, i.e., the HD is 50%. This is clearly observable in the average trend of HD from *MCNC* to *EPFL* to *ITC-99* benchmarks. Overall, the average HD for the respective benchmark suites is 36.5%, 40.02%, and 6.34%. For *EPFL* benchmarks presented in Fig. 7.11b, the HD numbers are less than those in the case of *MCNC* benchmarks. Among these, though the HD for *priority*

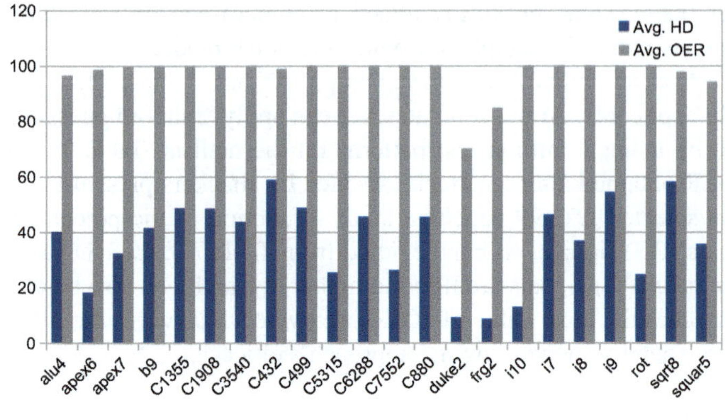

(a) MCNC

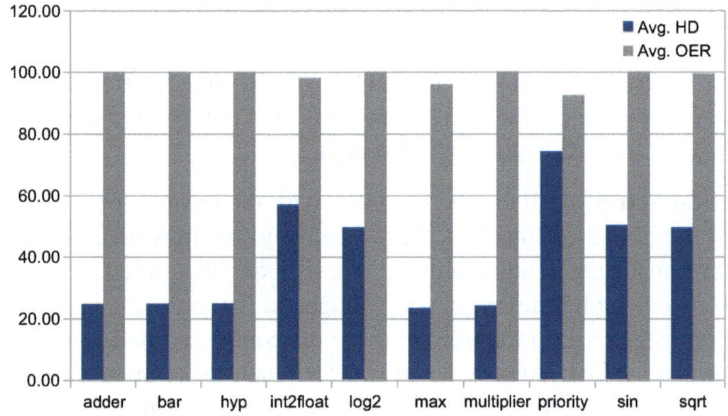

(b) EPFL

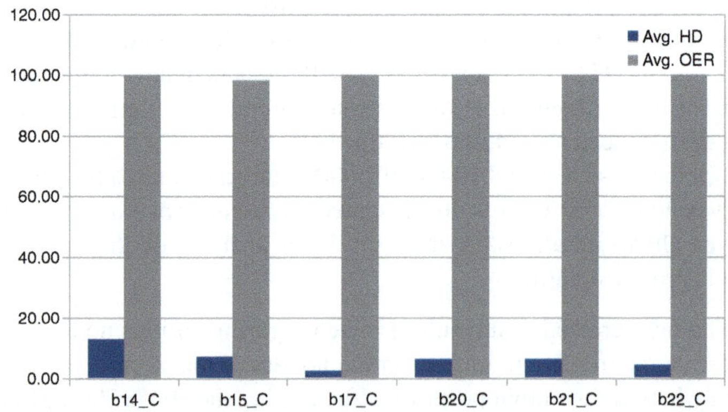

(c) ITC-99

Fig. 7.11 Evaluation of OER and HD to study the impact of voltage effects on the proposed security vulnerability. (**a**) Average Hamming distance and output error rate for *MCNC* benchmarks. (**b**) Average Hamming distance and output error rate for *EPFL* benchmarks. (**c**) Average Hamming distance and output error rate for *ITC-99* benchmarks is around 80%, it is equivalent to be around 20% if the inverted faulty output is converted as discussed above. Similarly, for the *ITC-99* benchmarks as shown in Fig. 7.11c, the average HD for the benchmarks is decreased as compared to both *MCNC* and *EPFL* benchmarks. As far as the OER is considered, for all the three benchmark suites, i.e., *MCNC*, *EPFL*, and *ITC-99* benchmarks, it mostly stabilizes at 99.99%, even with few misconfigurations.

It can be noticed that, as the size of the benchmarks increase, the HD stabilizes at lower values. However, an observation here is that there are exceptions, and the variations in the output bits suggest that *fault masking* plays an important role. *Fault masking* implies that the impact of a given fault in one part of the circuit may be masked due to another fault occurring at some other part of the circuit. The second reason can be attributed to the fact that, as the faults are inserted randomly, the distribution may not be uniform across the size of the benchmarks and hence can end up affecting only a small fraction of the total number of output ports. These are the reasons why *MCNC* benchmarks, especially *duke2*, a relatively small benchmark, demonstrates low HD. On the other hand, *frg2* and *i10*, which are reasonably large, show low HD due to masking effects. It is to be noted that this experiment only models the ambiguity in logic levels and demonstrates the impact of voltage effects caused due to the aforementioned security vulnerability. The model assumes that the incorrect configuration in RFET-based logic gate has already been introduced. As discussed earlier, current effects are stealthy and can lead to a far-reaching detrimental impact on the health of the host circuit.

7.8 Concluding Remarks and Future Research Directions

This work has highlighted the security promises and potential vulnerabilities in circuits based on emerging reconfigurable nanotechnologies. Design-for-security schemes such as logic locking and split manufacturing have been evaluated for RFET-based circuits. Using benchmark-level evaluations, these schemes established that transistor-level reconfigurability can provide effective security solutions as 100% OER and 31% HD is achieved for split manufacturing over *ITC-99* benchmarks.

Further, it has been demonstrated how security vulnerabilities could be exploited for RFET-based circuits using the very same transistor-level reconfigurability. Vulnerabilities occurring due to faults or misconfigurations of individual transistors have been evaluated using circuit-level simulations. The impact of *short-circuit* or *open-circuit* scenarios on representative circuits has been demonstrated. Both the *short-circuit* and the *open-circuit* configurations impact the current and the voltage levels and manifest in sequential as well as combinational circuits.

References

[Ade08] Sally Adee. "The hunt for the kill switch". In: *IEEE Spectrum* 45.5 (2008), pp. 34–39.

[AGD15] Luca Amarú, Pierre-Emmanuel Gaillardon, and Giovanni De Micheli. "The EPFL combinational benchmark suite". In: *Proceedings of the 24th International Workshop on Logic & Synthesis (IWLS)*. 2015.

[Bi+14] Y. Bi, P. E. Gaillardon, X. S. Hu, M. Niemier, J. S. Yuan, and Y. Jin. "Leveraging Emerging Technology for Hardware Security - Case Study on Silicon Nanowire FETs and Graphene SymFETs". In: *ATS*. 2014.

[Bi+16a] Yu Bi, X. Sharon Hu, Yier Jin, Michael Niemier, Kaveh Shamsi, and Xunzhao Yin. "Enhancing Hardware Security with Emerging Transistor Technologies". In: *Proceedings of the 26th Edition on Great Lakes Symposium on VLSI*. GLSVLSI '16. Boston, Massachusetts, USA: ACM, 2016, pp. 305–310. https://doi.org/10.1145/2902961.2903041.

[Bi+16b] Yu Bi, Kaveh Shamsi, Jiann-Shiun Yuan, Pierre-Emmanuel Gaillardon, Giovanni De Micheli, Xunzhao Yin, X Sharon Hu, Michael Niemier, and Yier Jin. "Emerging technology-based design of primitives for hardware security". In: *ACM Journal on Emerging Technologies in Computing Systems (JETC)* 13.1 (2016), pp. 1–19.

[Cha+20] A. Chakraborty, N. G. Jayasankaran, Y. Liu, J. Rajendran, O. Sinanoglu, A. Srivastava, Y. Xie, M. Yasin, and M. Zuzak. "Keynote: A Disquisition on Logic Locking". In: *IEEE Transactions on Computer- Aided Design of Integrated Circuits and Systems* 39.10 (2020), pp. 1952–1972. https://doi.org/10.1109/TCAD.2019.2944586.

[Che+16] A. Chen, X. S. Hu, Y. Jin, M. Niemier, and X. Yin. "Using emerging technologies for hardware security beyond PUFs". In: *2016 Design, Automation Test in Europe Conference Exhibition (DATE)*. Mar. 2016, pp. 1544–1549.

[Das14] Anup Kumar Das. "Design Methodologies for Reliable and Energy- Efficient Multiprocessor System". PhD thesis. 2014.

[Dom18] K. Domanski. "Latch-up in FinFET technologies". In: *2018 IEEE International Reliability Physics Symposium (IRPS)*. Mar. 2018, pp. 2C.4-1-2C.4–5. https://doi.org/10.1109/IRPS.2018.8353550.

[GG18] Edouard Giacomin and Pierre-Emmanuel Gaillardon. "Differential Power Analysis Mitigation Technique Using Three-Independent-Gate Field Effect Transistors". In: *IFIP/IEEE International Conference on Very Large Scale Integration (VLSI-SoC)*. 2018, pp. 107–112.

[Gor+19] G. Gore, P. Cadareanu, E. Giacomin, and P. Gaillardon. "A Predictive Process Design Kit for Three-Independent-Gate Field-Effect Transistors". In: *2019 IFIP/IEEE 27th International Conference on Very Large Scale Integration (VLSI-SoC)*. Oct. 2019, pp. 172–177. https://doi.org/10.1109/VLSI-SoC.2019.8920358.

[JKS87] Y. Ji-Ren, I. Karlsson, and C. Svensson. "A true single-phase-clock dynamic CMOS circuit technique". In: *IEEE Journal of Solid-State Circuits* 22.5 (Oct. 1987), pp. 899–901. https://doi.org/10.1109/JSSC.1987.1052831.

[Kim+14] Yoongu Kim, Ross Daly, Jeremie Kim, Chris Fallin, Ji Hye Lee, Donghyuk Lee, Chris Wilkerson, Konrad Lai, and Onur Mutlu. "Flipping bits in memory without accessing them: An experimental study of DRAM disturbance errors". In: *ACM SIGARCH Computer Architecture News* 42.3 (2014), pp. 361–372.

[Kne20] Johann Knechtel. "Hardware Security For and Beyond CMOS Technology: An Overview on Fundamentals, Applications, and Challenges". In: *Proceedings of the 2020 International Symposium on Physical Design*. ISPD '20. Taipei, Taiwan: Association for Computing Machinery, 2020, pp. 75–86. https://doi.org/10.1145/3372780.3378175.

[Koc+18] Paul Kocher, Jann Horn, Anders Fogh, Daniel Genkin, Daniel Gruss, Werner Haas, Mike Hamburg, Moritz Lipp, Stefan Mangard, Thomas Prescher, Michael Schwarz, and Yuval Yarom. "Spectre Attacks: Exploiting Speculative Execution". In: *arXiv abs/ 1801.01203* (2018). arXiv: 1801.01203.

[KPS19] J. Knechtel, Satwik Patnaik, and Ozgur Sinanoglu. "Protect Your Chip Design Intellectual Property: An Overview". In: *Proc. Int. Conf. Omni-Layer Intelligent Systems (COINS)*. 2019, pp. 211–216. https://doi.org/10.1145/3312614.3312657.

[LH96] Hyung Ki Lee and Dong Sam Ha. "HOPE: an efficient parallel fault simulator for synchronous sequential circuits". In: *TCAD* (1996).

[Lip+18] Moritz Lipp, Michael Schwarz, Daniel Gruss, Thomas Prescher, Werner Haas, Stefan Mangard, Paul Kocher, Daniel Genkin, Yuval Yarom, and Mike Hamburg. "Meltdown". In: *ArXiv e-prints* (Jan. 2018). arXiv: 1801.01207.

[Mar+12] M. De Marchi, D. Sacchetto, S. Frache, J. Zhang, P. E. Gaillardon, Y. Leblebici, and G. De Micheli. "Polarity control in double-gate, gate-all-around vertically stacked silicon nanowire FETs". In: *2012 International Electron Devices Meeting*. Dec. 2012, pp. 8.4.1–8.4.4. https://doi.org/10.1109/IEDM.2012.6479004.

[McD+16] J. T. McDonald, Y. C. Kim, T. R. Andel, M. A. Forbes, and J. McVicar. "Functional polymorphism for intellectual property protection". In: *HOST*. 2016.

[Mik+17] T Mikolajick, A Heinzig, J Trommer, T Baldauf, and W M Weber. "The RFET—a reconfigurable nanowire transistor and its application to novel electronic circuits and systems". In: *Semiconductor Science and Technology* 32.4 (2017), p. 043001.

[Pat+18a] S. Patnaik, N. Rangarajan, J. Knechtel, O. Sinanoglu, and S. Rakheja. "Advancing Hardware Security Using Polymorphic and Stochastic Spin-Hall Effect Devices". In: *Proc. DATE*. 2018, pp. 97–102. https://doi.org/10.23919/DATE.2018.8341986.

[Pat+18b] Satwik Patnaik, Mohammed Ashraf, Johann Knechtel, and Ozgur Sinanoglu. "Raise your game for split manufacturing: Restoring the true functionality through BEOL". In: *55th Design Automation Conference (DAC)*. 2018, pp. 1–6.

[Pat+18c] Satwik Patnaik, Johann Knechtel, Mohammed Ashraf, and Ozgur Sinanoglu. "Concerted wire lifting: Enabling secure and cost-effective split manufacturing". In: *23rd Asia and South Pacific Design Automation Conference (ASP-DAC)*. 2018, pp. 251–258.

[Rai+19a] S. Rai, A. Rupani, P. Nath, and A. Kumar. "Hardware Watermarking Using Polymorphic Inverter Designs Based On Reconfigurable Nanotechnologies". In: *ISVLSI*. 2019.

[Rai+19b] S. Rai, J. Trommer, M. Raitza, T. Mikolajick, W. M. Weber, and A. Kumar. "Designing Efficient Circuits Based on Runtime-Reconfigurable Field-Effect Transistors". In: *IEEE Transactions on Very Large Scale Integration (VLSI) Systems* 27.3 (Mar. 2019), pp. 560–572. https://doi.org/10.1109/TVLSI.2018.2884646.

[Raj+15] J. Rajendran, R. Karri, J. B. Wendt, M. Potkonjak, N. McDonald, G. S. Rose, and B. Wysocki. "Nano Meets Security: Exploring Nanoelectronic Devices for Security Applications". In: *Proceedings of the IEEE* 103.5 (May 2015), pp. 829–849. https://doi.org/10.1109/JPROC.2014.2387353.

[RKK14] M. Rostami, F. Koushanfar, and R. Karri. "A Primer on Hardware Security: Models, Methods, and Metrics". In: *Proceedings of the IEEE* (2014).

[RRK18] S. Rai, M. Raitza, and A. Kumar. "Technology mapping flow for emerging reconfigurable silicon nanowire transistors". In: *2018 Design, Automation Test in Europe Conference Exhibition (DATE)*. Mar. 2018, pp. 767–772. https://doi.org/10.23919/DATE.2018.8342110.

[RRK19] A. Rupani, S. Rai, and A. Kumar. "Exploiting Emerging Reconfigurable Technologies for Secure Devices". In: *Euromicro DSD*. 2019.

[Sed+14] Behnam Sedighi, Xiaobo Sharon Hu, Joseph J. Nahas, and Michael Niemier. "Nontraditional Computation Using Beyond-CMOS Tunneling Devices". In: *IEEE Journal on Emerging and Selected Topics in Circuits and Systems* 4.4 (2014), pp. 438–449. https://doi.org/10.1109/JETCAS.2014.2361065.

[SEM12] SEMI. *Innovation is at risk as semiconductor equipment and materials industry loses up to $4 billion annually due to IP infringement*. 2012.

[Sen+19] Abhrajit Sengupta, Mohammed Nabeel, Johann Knechtel, and Ozgur Sinanoglu. "A New Paradigm in Split Manufacturing: Lock the FEOL, Unlock at the BEOL". In: *Design, Automation & Test in Europe Conference & Exhibition (DATE)*. 2019, pp. 414–419.

[Sha+17] Kaveh Shamsi, Meng Li, Travis Meade, Zheng Zhao, David Z Pan, and Yier Jin. "AppSAT: Approximately deobfuscating integrated circuits". In: *IEEE International Symposium on Hardware Oriented Security and Trust (HOST)*. 2017, pp. 95–100.

[Sha+20] Mohammad Mehdi Sharifi, Ramin Rajaei, Patsy Cadareanu, Pierre- Emmanuel Gaillardon, Yier Jin, Michael Niemier, and X Sharon Hu. "A novel TIGFET-based DFF design for improved resilience to power side-channel attacks". In: *Proceedings of the 23rd Conference on Design, Automation and Test in Europe*. 2020, pp. 1253–1258.

[Shi+10] Yuriy Shiyanovskii, F Wolff, Aravind Rajendran, C Papachristou, D Weyer, and W Clay. "Process reliability based trojans through NBTI and HCI effects". In: *2010 NASA/ESA Conference on Adaptive Hardware and Systems*. IEEE. 2010, pp. 215–222.

[SR14] V. Simek and R. Ruzicka. "Reconfigurable Platform with Polymorphic Digital Gates and Partial Reconfiguration Feature". In: *European Modelling Symposium*. 2014.

[SRM15] P. Subramanyan, S. Ray, and S. Malik. "Evaluating the security of logic encryption algorithms". In: *Proc. HOST*. 2015, pp. 137–143. https://doi.org/10.1109/HST.2015.7140252.

[Svi+10] Vjekoslav Svilan, Tien-Min Chen, Kleanthes G Koniaris, and James B Burr. *Method and system for latchup suppression*. US Patent 7,786,756. Aug. 2010.

[Tro+15] Jens Trommer, André Heinzig, Tim Baldauf, Stefan Slesazeck, Thomas Mikolajick, and Walter M. Weber. "Functionality-Enhanced Logic Gate Design Enabled by Symmetrical Reconfigurable Silicon Nanowire Transistors". In: *IEEE Transactions on Nanotechnology* 14.4 (July 2015), pp. 689–698. https://doi.org/10.1109/TNANO.2015.2429893.

[Wan+18] Y. Wang, P. Chen, J. Hu, G. Li, and J. Rajendran. "The Cat and Mouse in Split Manufacturing". In: *IEEE Transactions on Very Large Scale Integration (VLSI) Systems* 26.5 (2018), pp. 805–817. https://doi.org/10.1109/TVLSI.2017.2787754.

[Xia+16] Kan Xiao, Domenic Forte, Yier Jin, Ramesh Karri, Swarup Bhunia, and Mohammad Tehranipoor. "Hardware trojans: Lessons learned after one decade of research". In: *ACM Transactions on Design Automation of Electronic Systems (TODAES)* 22.1 (2016), pp. 1–23.

[Yas+19] Muhammad Yasin, Bodhisatwa Mazumdar, Jeyavijayan Rajendran, and Ozgur Sinanoglu. "Hardware Security and Trust: Logic Locking as a Design-for-Trust Solution". In: *The IoT Physical Layer: Design and Implementation*. Ed. by Ibrahim (Abe) M. Elfadel and Mohammed Ismail. Cham: Springer International Publishing, 2019, pp. 353–373. https://doi.org/10.1007/978-3-319-93100-5_20.

[Zha+14] Jian Zhang, Xifan Tang, Pierre Emmanuel Gaillardon, and Giovanni De Micheli. "Configurable Circuits Featuring Dual-Threshold-Voltage Design With Three-Independent-Gate Silicon Nanowire FETs". In: *IEEE Transactions on Circuits and Systems I: Regular Papers* 61.10 (Oct. 2014), pp. 2851–2861. https://doi.org/10.1109/TCSI.2014.2333675.

Chapter 8
Conclusion

In this book, a custom bottom-up approach towards enabling Electronic Design Automation (EDA) for circuits based on Reconfigurable Field-Effect Transistor (RFET) has been proposed. Towards this end, this chapter presents the concluding remarks, important takeaways, and directions for future work.

8.1 Concluding Remarks

The ability to reconfigure the most atomic entity in an electronic circuit, i.e., the transistor, has opened up new possibilities in terms of circuit designs and meeting application requirements. Emerging reconfigurable nanotechnology offers (re)programming of the transistors to either behave as a p- or n-type device. The (re)programmability is only possible since Reconfigurable Field-Effect Transistors (RFETs) demonstrate near-to-full electrical symmetry between the two configurations. Thus, reconfigurable nanotechnology has blurred the very notion of separation between the two types of devices—p-channel Field-Effect Transistor (PFET) and n-channel Field-Effect Transistor (NFET)—which has been the characteristic of the contemporary CMOS technology. Consequently, the two configurations of a single RFET exhibit equal current drives. This, in turn, provides the flexibility to have a complementary circuit design, with equal drive strengths for both pull-up and pull-down networks. Exploiting this flexibility, the book presents a range of RFET-based logic gates and circuits capable of exhibiting enhanced functionality. This is necessary to complement CMOS technology downscaling from a functional point of view.

Moreover, the reconfigurable properties are not only limited to displaying electrical symmetry but to also provide additional flexibility such as having multiple-independent gate terminals and altering the performance of individual FETs by steering potential at different gate terminals. The book proposes various approaches

S. Rai, A. Kumar, *Design Automation and Applications for Emerging Reconfigurable Nanotechnologies*, https://doi.org/10.1007/978-3-031-37924-6_8

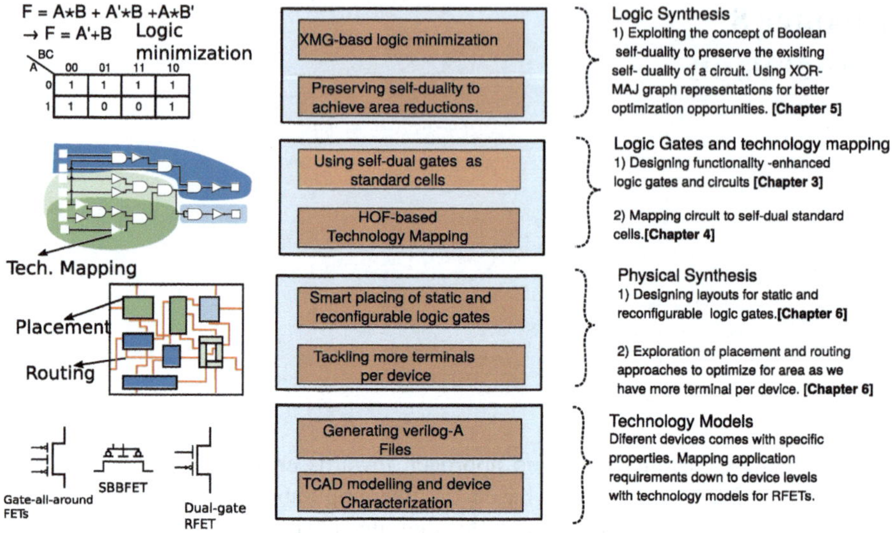

Fig. 8.1 Book correlation and important contributions towards enabling EDA flow for RFET-based circuits

to unlock these potential by offering contributions to the whole EDA flow. These contributions include developing both combinational and sequential circuit design paradigms, proposing technology-mapping flows, designing standard cells based on RFETs, developing first-ever technology models for carrying out physical synthesis of RFET-based circuits, and finally, exploiting RFET's polymorphic behavior to implement affordable security measures.

The book follows a close correlation with the various stages of an EDA flow. Hence, each subsequent chapter (Chaps. 3, 4, 5, and 6) caters to the stages of an EDA flow as shown in Fig. 8.1. The figure summarizes the book's main contribution to various stages of EDA as introduced in Chap. 1. For the ease of understanding, the left portion of Fig. 8.1 is the same as in Chap. 1. The contributions are aimed towards the main research objective mentioned in Chap. 1, i.e., to develop an EDA flow to enable RFET-based circuits. Additionally, it has been shown in Chap. 7 that RFETs-based primitives can play a major role in securing circuits. At the time of writing this book, the motivation to provide secure circuits at low overheads is one of the biggest drivers in enabling RFET integration into Integrated Circuitss (ICs). Targeted towards enabling an Electronic Design Automation (EDA) flow, this book aims to develop a custom yet CMOS-compatible EDA flow for RFETs for designing efficient and secure circuits. The prime reason why CMOS-compatible flow was targeted is that RFETs can be readily adopted with minimal overheads in terms of mainstream integration. The approaches and the flows presented in the book are consistent with the typical CMOS-centric EDA flows, albeit with features supporting RFET's integration.

Interestingly, there are various contending technologies based on materials such as silicon and germanium that can realize reconfigurable FETs. As we saw in Chap. 2, devices for both 1D and 2D device geometry have been demonstrated with reconfigurable properties. Most of these technologies are based on materials that are predominantly used in conventional CMOS. Several of these technologies share the manufacturing process stack of CMOS and are based on transistors' geometries that are CMOS-compatible.

Motivated by the flexibility offered by RFETs, circuit designs proposed in Chap. 3 demonstrate that CMOS-styled circuit designs are sub-optimal for RFETs. They do provide a good starting point to design circuits based on RFETs but fail to capitalize on the reconfigurable properties of transistors. Both sequential and combinational circuit designs employed either the flexibility offered via multi-independent gates or variable performance for efficient implementation based on RFETs. Hence, various functionality-enhanced logic gate designs have been demonstrated with better area and performance as compared to the CMOS designs. One major takeaway from these circuit designs is that the NAND, NOR, and multiplexer implementation all have the same current drive as an inverter. This is truly disruptive as it facilitates the packing of more functionality with no performance overhead. Further, the feature to support multi-independent gate terminals enables logic gate designs for logic functions with more than two inputs which were considered complex in CMOS due to cascading of multiple transistors and due to sizing issues. Thus, efficient circuit designs for 3-XOR and MINORITY are proposed which use the reconfigurability offered by individual transistors. The same transistor-level reconfigurability is utilized to make the quintessential True-Single Phase Clock (TSPC) D Flip-Flop (DFF) as dual-edge triggered.

Based on these functionality-enhanced logic gates, two major research directions were explored in Chap. 4. First, the Boolean property of self-duality has been proposed as the natural logical abstraction for reconfigurable nanotechnology. It was shown that the self-dual Boolean logic functions can truly tap the duality of functionality at the transistor level. Thus, self-dual Boolean logic functions can be implemented with RFETs by switching between the pull-up and pull-down networks. Using the property of self-duality, an algorithmic approach has been presented to distill standard cells based on RFETs from a given benchmark suite. The major takeaway is that efficient standard cells based on self-dual Boolean logic can lead to area reduction for RFET-based circuits. Additionally, Chap. 4 also proposed a technology mapping flow for mutually exclusive functions based on the concept of Higher-Order Function (HOF). The support for such technology mapping has been integrated within the ABC tool [BM10]. Using this technology mapper, an early evaluation was presented to demonstrate the overheads in terms of the number of transistors and the delay when using reconfigurable functionality-enhanced logic gates.

The takeaway from Chap. 4 has been the biggest motivation to enable a tailor-made logic synthesis flow for RFETs. Since self-dual logic functions are implemented efficiently using RFETs, Chap. 5 presented a logic synthesis flow with the aim to preserve the maximum available self-duality in a given circuit. Preserving self-duality is important since logic synthesis and technology mapping stages depend heavily on cut-based optimization which can disrupt the existing self-duality of a circuit. Hence, XMG-based logic representation has been used during the logic synthesis and technology mapping since they offer compact representation for both unate and binate functions. Advanced Boolean methods such as resubstitution and rewriting were proposed for XMGs to preserve and increase the existing self-duality. Using a comprehensive analysis, it has been demonstrated that the XMG-based approach offers better area results as compared to AIGs. The contributions in this chapter again instantiate that CMOS-style flows may be sub-optimal for RFET-based circuits.

While the first three chapters focused on developing circuit designs and enabling logic synthesis flows, Chap. 6 presented a physical synthesis flow for RFET-based circuits. An entire EDA flow has been presented that takes a given circuit description as an input and outputs an actual physical layout for circuits based on RFETs exclusively. Using the technology model for an established Silicon Nanowire (SiNW) RFET model [Bla+17], liberty characterization files and optimized layouts for both static and functionality-enhanced logic gates have been proposed. Experimental evaluations demonstrate that RFET-based implementations have higher area overheads as compared to CMOS-based implementations. Considering that the liberty models were one of the first proposed models for RFETs, this evaluation was significant and fostered research directions to improve the RFET models.

The RFET technology, albeit in its infancy, has shown a lot of promise in showing better numbers for area and delay for its functionality-enhanced logic gates. These logic gates and circuit design paradigms naturally complement the demands of hardware security. Owing to its inherent transistor-level reconfigurability, it was demonstrated in Chap. 7 that security measures based on RFETs can be implemented in an effective and cost-efficient manner. Both logic locking and split-manufacturing schemes proposed using RFETs demonstrated low overheads in terms of area and delay. As compared to CMOS-based schemes that demand insertion of extra logic gates for implementing these two security schemes, RFET's transistor-level reconfiguration realizes these schemes without the need for the insertion of additional logic. However, it has also been demonstrated in Chap. 7 that the same transistor-level reconfiguration can lead to circuit vulnerabilities in the case of malicious misconfiguration of an individual or group of RFETs in a circuit. These vulnerabilities are detrimental as such misconfigurations can manifest as a short-circuit or an open-circuit condition and can lead to partial or complete impairment of the chip.

8.2 Directions for Future Work

While this book presented an entire EDA flow for enabling RFET-based circuits, there are certain research directions that remain to be explored.

1. *Exploration of newer technology models for RFETs:* The disruptive reconfigurable technology is not limited to silicon and is expandable to other semiconductor materials, like germanium [Tro17] and carbon [OCo+12, Bla+17]. Several of these technologies share the manufacturing process stack of CMOS. Newer technology solutions delivering a more compact structure of RFETs, such as vertical or stacked technologies, can play a major role in further reducing the area for circuits based on RFETs. For example, germanium is already used in p-channels in contemporary CMOS technology to boost performance [Car+16]. Preliminary evaluations using Germanium Nanowire (GeNW) show promising results in terms of better performance as compared to the conventional silicon-based RFETs. The stacking of nanowires or nanosheets can also play a major role in improving the device's drive strength.
2. *Circuit design:* A majority of research in RFETs is focused from a digital design perspective [Rai+17, Gai+14]. Barring few works [Gai+16, NKS16], custom circuit design based on RFETs is still, rather, unexplored. The unique properties of RFETs can be used for designing efficient analog circuit designs. Similarly, sequential logic designs such as flip-flops and latches can also benefit from the dynamic reconfiguration of electrical conduction of individual transistors. Similarly, multi-V_t application scenarios should be explored because unlike CMOS, where transistors with variable threshold voltages are fixed, RFETs allow independent reconfiguration of individual transistors by steering the potential at the program and control gates.
3. *Logic synthesis:* The main limitation of the current logic synthesis flow is that it gives inferior results as compared to AIGs for circuits with lower self-duality. In this direction, better heuristics need to be proposed where logic synthesis can utilize the benefits of both AIGs and XMGs within the same logic graph. This will help in having a uniform flow for all kinds of RFET-based circuits. Exploring other Boolean methods such as Exclusive Sum-of-Products (ESOP) balancing of the graph structure and refactoring using XMGs is an interesting research direction.
4. *Standard cells and technology mapping:* In Chap. 4, the algorithm for calculating area savings due to inverter adjustments can be improved. Presently, it is done as a step post mapping. However, it can be done during technology mapping as well. Better heuristics are required to preserve and optimize self-dual portions of the circuit during the technology mapping stage since we have seen that more self-dual cuts are possible with XMGs than AIGs. Hence, exploring custom heuristics can be an interesting research direction. Recently, model-checking-based work [Rai+22] demonstrates that various designs of standard cells can be implemented using RFETs for different parameters such as average delay, maximum delay, etc. Integration of these designs into mainstream technology mapping can be worth exploring.
5. *Physical synthesis:* To further improve physical synthesis for reconfigurable circuits, possibilities for more efficient generation of $P/\overline{P}$ signals should be explored. However, such endeavors require better layouts and specialized PDKs for logic gates based on RFETs. This, in turn, depends upon better RFET

technology models as mentioned in Point 1. Furthermore, measures like all-around devices and stacking of nanowires [Mer+16, Ern+06] will lead to a substantial reduction of device width and accordingly to a reduced cell area of the layouts. This can impact the overall placement of individual standard cells and hence the routing. A high number of metal layers and a small pitch between gate terminals can further relax routing constraints in order to facilitate physical synthesis for RFETs. Similarly, the development of libraries for multi-independent gate RFETs (MIGFETs or TIGFETs) offers the potential to improve the overall post-physical synthesis of RFET-based circuits. The greatest promise that SiNWs hold is that they are compatible with the current silicon technology. This opens up new avenues to explore the feasibility of making SiNW-CMOS complementary hybrid layouts.

6. *Exploring security schemes:* This work aims to open up interesting future research directions regarding prospects pertaining to hardware security (in terms of both security promises and security vulnerabilities), which can accelerate the commercial integration of RFETs in electronic circuits. Particularly, security vulnerabilities are more detrimental as the electrical repercussions caused by such a scenario are far more severe as they can present themselves in the form of metastability, change of critical paths, and higher dynamic and static power dissipation.

   RFETs, in general, have the potential to be a game-changer technology in enabling hardware security. Future work directions include exploring diverse applications such as tapping of metastability for physically unclonable functions (PUFs) [Che+16] and random number generators [Bha+21]. Exploring analog domains with RFETs can also be crucial in defining security implications at the IC level. While the preliminary investigation in RFET-based circuits demonstrates robustness against various side-channel attacks due to its symmetric electrical characteristics in both p- and n-type behavior, detailed experimental evaluations involving differential analysis are needed to provide security guarantees.
7. *Exploration of other application:* Apart from hardware security, RFET's unique properties can be utilized in various other applications. One promising direction of research is to explore reconfigurable computing fabrics based on RFETs. There have been a few works in this direction where fine-grained architectures based on RFETs have been proposed [Gai+15, CBO17]. RFET's reconfigurable properties can be used in architectures such as Coarse-Grained Reconfigurable Array (CGRA) or Field-programmable Gate Array (FPGA). RFETs can also be used as a programming device for nonvolatile memories with possible co-integration of FeFETs with RFETs [Rai+21]. This nanotechnology provides the opportunity to switch its conduction on the fly from p-type to n-type in a nonvolatile fashion. Interesting circuit design opportunities can be explored with nonvolatility in conjunction with reconfiguration.

Overall, the book advances RFETs towards commercial feasibility by proposing various approaches to enable design automation for efficient and secure RFET-based circuits. Besides, the contributions made in this book aim to foster new research paths in enabling EDA flow and exploring various applications for emerging reconfigurable technologies.

References

[Bha+21] Abhiroop Bhattacharjee, Shubham Rai, Ansh Rupani, Michael Raitza, and Akash Kumar. "Metastability with Emerging Reconfigurable Transistors: Exploiting Ambipolarity for Throughput". In: *2021 IFIP/IEEE 29th International Conference on Very Large Scale Integration (VLSI-SoC)*. 2021, pp. 1–6. https://doi.org/10.1109/VLSI-SoC53125.2021.9607015.

[Bla+17] S. Blawid, D. L. M. de Andrade, S. Mothes, and M. Claus. "Performance Projections for a Reconfigurable Tunnel NanoFET". In: *IEEE Journal of the Electron Devices Society* 5.6 (Nov. 2017), pp. 473–479. https://doi.org/10.1109/JEDS.2017.2756040.

[BM10] Robert Brayton and Alan Mishchenko. "ABC: An academic industrialstrength verification tool". In: *Computer Aided Verification*. Springer. 2010.

[Car+16] R. Carter, J. Mazurier, L. Pirro, J-U. Sachse, P. Baars, J. Faul, C. Grass, G. Grasshoff, P. Javorka, T. Kammler, A. Preusse, S. Nielsen, T. Heller, J. Schmidt, H. Niebojewski, P-Y. Chou, E. Smith, E. Erben, C. Metze, C. Bao, Y. Andee, I. Aydin, S. Morvan, J. Bernard, E. Bourjot, T. Feudel, D. Harame, R. Nelluri, H.-J. Thees, L. M-Meskamp, J. Kluth, R. Mulfinger, M. Rashed, R. Taylor, C. Weintraub, J. Hoentschel, M. Vinet, J. Schaeffer, and B. Rice. "22nm FDSOI technology for emerging mobile, Internet-of-Things, and RF applications". In: *2016 IEEE International Electron Devices Meeting (IEDM)*. 2016, pp. 2.2.1–2.2.4. https://doi.org/10.1109/IEDM.2016.7838029.

[CBO17] K. Cheng, S. Le Beux, and I. O'Connor. "Hybrid Topologies for Reconfigurable Matrices Based on Nano-Grain Cells". In: *2017 IEEE International Conference on Rebooting Computing (ICRC)*. Nov. 2017, pp. 1–8. https://doi.org/10.1109/ICRC.2017.8123639.

[Che+16] A. Chen, X. S. Hu, Y. Jin, M. Niemier, and X. Yin. "Using emerging technologies for hardware security beyond PUFs". In: *2016 Design, Automation Test in Europe Conference Exhibition (DATE)*. Mar. 2016, pp. 1544–1549.

[Ern+06] T. Ernst, C. Dupre, C. Isheden, E. Bernard, R. Ritzenthaler, V. Maffini-Alvaro, J. C. Barbe, F. De Crecy, A. Toffoli, C. Vizioz, S. Borel, F. Andrieu, V. Delaye, D. Lafond, G. Rabille, J. M. Hartmann, M. Rivoire, B. Guillaumot, A. Suhm, P. Rivallin, O. Faynot, G. Ghibaudo, and S. Deleonibus. "Novel 3D integration process for highly scalable Nano-Beam stacked-channels GAA (NBG) FinFETs with HfO2/TiN gate stack". In: *2006 International Electron Devices Meeting*. Dec. 2006, pp. 1–4. https://doi.org/10.1109/IEDM.2006.346955.

[Gai+14] P. E. Gaillardon, L. Amaru, J. Zhang, and G. De Micheli. "Advanced system on a chip design based on controllable-polarity FETs". In: *2014 Design, Automation Test in Europe Conference Exhibition (DATE)*. Mar. 2014, pp. 1–6. https://doi.org/10.7873/DATE.2014.248.

[Gai+15] Pierre Emmanuel Gaillardon, Xifan Tang, Gain Kim, and Giovanni De Micheli. "A Novel FPGA Architecture Based on Ultrafine Grain Reconfigurable Logic Cells". In: *IEEE Transactions on Very Large Scale Integration (VLSI) Systems* 23.10 (Oct. 2015), pp. 2187–2197. DOI:10.1109/TVLSI.2014.2359385.

[Gai+16] Pierre-Emmanuel Gaillardon, Mehdi Hasan, Anirban Saha, Luca Amarú, RossWalker, and Berardi Sensale Rodriguez. "Digital, analog and RF design opportunities of three-independent-gate transistors". In: *2016 IEEE International Symposium on Circuits and Systems (ISCAS)*. 2016, pp. 405–408. https://doi.org/10.1109/ISCAS.2016.7527256.

[Mer+16] H. Mertens, R. Ritzenthaler, A. Chasin, T. Schram, E. Kunnen, A. Hikavyy, L. Å. Ragnarsson, H. Dekkers, T. Hopf, K. Wostyn, K. Devriendt, S. A. Chew, M. S. Kim, Y. Kikuchi, E. Rosseel, G. Mannaert, S. Kubicek, S. Demuynck, A. Dangol, N. Bosman, J. Geypen, P. Carolan, H. Bender, K. Barla, N. Horiguchi, and D. Mocuta. "Vertically stacked gate-all-around Si nanowire CMOS transistors with dual work function metal gates". In: *2016 IEEE International Electron Devices Meeting (IEDM)*. Dec. 2016, pp. 19.7.1–19.7.4. https://doi.org/10.1109/IEDM.2016.7838456.

[NKS16] Kaushal Nigam, Pravin Kondekar, and Dheeraj Sharma. "DC characteristics and analog/RF performance of novel polarity control GaAs-Ge based tunnel field effect transistor". In: *Superlattices and Microstructures* 92 (2016), pp. 224–231.

[OCo+12] Ian O'Connor, Kotb Jabeur, Sébastien Le Beux, and David Navarro. "Ambipolar Independent Double Gate FET Logic". In: *Proceedings of the 2012 IEEE/ACM International Symposium on Nanoscale Architectures*. NANOARCH '12. Amsterdam, The Netherlands: ACM, 2012, pp. 61–68. https://doi.org/10.1145/2765491.2765504.

[Rai+17] M. Raitza, A. Kumar, M. Völp, D.Walter, J. Trommer, T. Mikolajick, and W. M. Weber. "Exploiting transistor-level reconfiguration to optimize combinational circuits". In: *Design, Automation Test in Europe Conference Exhibition (DATE), 2017*. Mar. 2017, pp. 338–343. https://doi.org/10.23919/DATE.2017.7927013.

[Rai+21] Shubham Rai, Mengyun Liu, Anteneh Gebregiorgis, Debjyoti Bhattacharjee, Krishnendu Chakrabarty, Said Hamdioui, Anupam Chattopadhyay, Jens Trommer, and Akash Kumar. "Perspectives on Emerging Computation-in-Memory Paradigms". In: *2021 Design, Automation Test in Europe Conference Exhibition (DATE)*. 2021, pp. 1925–1934. https://doi.org/10.23919/DATE51398.2021.9473976.

[Rai+22] Michael Raitza, Steffen Märcker, Shubham Rai, and Akash Kumar. "Exploring Standard-Cell Designs for Reconfigurable Nanotechnologies: A Formal Approach". In: *2022 Design, Automation Test in Europe Conference Exhibition (DATE)*. Mar. 2022.

[Tro17] Jens Trommer. *Towards Reconfigurable Electronics by Functionality- Enhanced Circuits and Germanium Nanowire Devices*. BoD–Books on Demand, 2017.

Appendix A
RFET-Based Genlib

Table A.1 Contents of the generic library for RFETs

Name of the gate	Identifiers	Nos. of fn	Area	1st fn (HOF)	2nd fn, 3rd fn
inv1			1	O=!a	
nand2_ nor2_ min3	sinw	3	2.625	O1=((!(a*b)) * !c) + ((!(a+b)) * c)	O2 =!(a*b), O3 = !(a_b)
xor2_ xnor2_ xor3f	sinw inv_adj	3	5.5	O1=((a ⊕ b) * !p) + (!(a ⊕ b) * p)	O2=a ⊕ b O3= !(a⊕ b)
mux			4.5	O=(!s*a)+(b*s)	
aoi_oai21	sinw	2	4	O1=!(a*b+c)	O2=!((a+b)*c)
emux			8.5	O=((!(a*c + (!c)*b))*e) + (!(a*d + (!d)*b)*(!e))	
nand3_nor3	sinw	2	3.25	O1= !(a*b*c)	O2 = !(a+b+c)
maj_min	sinw	2	4.75	O1= ((a*b)+(b*c) +(c*a))	O2=!((a*b)+(b*c)+(c*a))

S. Rai, A. Kumar, *Design Automation and Applications for Emerging Reconfigurable Nanotechnologies*, https://doi.org/10.1007/978-3-031-37924-6

Appendix A

[illegible]

Appendix B
Distilling Standard Cells

Table B.1 Occurrence of self-dual functions in the EPFL benchmark suite

S.no.	Number of inputs	Truth table	Number of occurrences
1	3	01101001	138400
2	3	00010111	34011
3	5	10010110011010010110100110010110	29626
4	3	01110001	21380
5	3	11010100	10914
6	3	10110010	8672
7	5	00010111111010001110100000010111	1876
8	5	10101001100101010101011001101010	1388
9	5	11010100001010110010101111010100	1067
10	5	10001110011100010111000110001110	891
11	5	01001101101100101011001001001101	793
12	4	0000011010011111	470
13	5	00000000000101110001011111111111	364
14	5	11101000111111110000000011101000	321
15	4	0110000011111001	314
16	5	10010101101010010110101001010110	212
17	4	0011011100010011	212
18	5	11001001100100110011011001101100	190
19	5	10011010010110010110010110100110	184
20	5	11111111000101110001011100000000	182
21	5	01011001100110101010011001100101	174
22	5	01101001111111110000000001101001	146

(continued)

S. Rai, A. Kumar, *Design Automation and Applications for Emerging Reconfigurable Nanotechnologies*, https://doi.org/10.1007/978-3-031-37924-6

Table B.1 (continued)

S.no.	Number of inputs	Truth table	Number of occurrences
23	5	10010011011011001100100100110110	125
24	5	000100110111111110000000100110111	125
25	4	1011010011010010	120
26	4	0000100101101111	104
27	4	1001000011110110	104
28	5	11111110111010101010100010000000	94
29	5	11110111001100010111001100010000	80
30	5	11111111100101101001011000000000	67
31	4	0111111010000001	66
32	5	11000111001110001110001100011100	59
33	5	11101011100000101011111000101000	59
34	4	0110010101011001	58
35	5	000001110011111110000001100011111	58
36	4	1110110011001000	58
37	5	101111110001010001110101110000010	55
38	4	0101100101100101	50
39	4	0100001010111101	48
40	4	0110101010101001	42
41	5	11111111101100101011001000000000	38
42	5	11111110110100101101001000000000	35
43	5	000101111111111100000000000010111	34
44	5	100101101111111110000000010010110	32
45	5	11100001100001110001111001111000	31
46	5	100011101111111110000000010001110	30
47	4	1000000011111110	24
48	4	1101010101010100	24
49	5	10011001100101101001011001100110	24
50	5	01100101011010010110100101011001	22
51	5	00111001100111001100011001100011	17
52	5	11111110111000101110001000000000	17
53	4	0000110101001111	16
54	4	0110110011001001	16
55	4	0001100011100111	14
56	5	1011001011111111100000000010110010	12
57	4	1110101010101000	12
58	4	0011000101110011	12
59	5	11111110010101100101011000000000	12
60	5	11111111100011101000111000000000	11
61	5	01011010011010010110100110100101	11
62	5	10010011110010010110110000110110	10
63	4	0010010011011011	10
64	4	1011001100110010	10
65	4	0001101100100111	10
66	5	00111100011010010110100111000011	8

(continued)

Table B.1 (continued)

S.no.	Number of inputs	Truth table	Number of occurrences
67	4	0001111100000111	8
68	4	0101011010010101	8
69	4	0101011100010101	8
70	5	10001110000000001111111110001110	8
71	5	11101111101011101000101000001000	8
72	5	10000010001010001110101110111110	8
73	4	1111100011100000	8
74	5	00000100010011001100110111011111	7
75	5	11111011101010101010100010000000	7
76	5	11101010111111101000000010101000	6
77	4	0101000101110101	6
78	5	10101110111011110000100010001010	6
79	5	01101001100110010110011001101001	6
80	4	0001001010110111	6
81	5	10110010000000001111111110110010	6
82	5	10111110111010110010100010000010	6
83	4	1100100101101100	6
84	5	10011100001110010110001111000110	5
85	5	10101011101111111000000100010101 0	5
86	5	11111111010011010100110100000000	4
87	5	10101000100000001111111011101010	4
88	5	10110010001010110010101110110010	4
89	5	11111111110101001101010000000000	4
90	5	10111111101010110010101000000010	4
91	5	11001001001101101001001101101100	4
92	4	0010110101001011	4
93	4	1100011010011100	4
94	4	1000111100001110	4
95	5	00101011000000001111111100101011	4
96	5	00101011111111110000000000101011	4
97	4	1000011100011110	4
98	4	0111001100110001	4
99	4	0101010011010101	4
100	4	0011101100100011	4
101	5	01110011000100001111011100110001	4
102	5	10010000000010010110111111110110	3
103	5	11011101010011010100110101000100	3
104	4	0100111100001101	2
105	4	0010111100001011	2
106	4	1010001010111010	2
107	4	1011111100000010	2
108	5	10101001010101101001010101101010	2
109	4	1100110101001100	2
110	5	11110001111101110001000001110000	2

(continued)

Table B.1 (continued)

S.no.	Number of inputs	Truth table	Number of occurrences
111	4	1101011100010100	2
112	4	1101110011000100	2
113	4	1111000101110000	2
114	5	11010101111111010100000001010100	2
115	5	01010010010010101010110110110101	2
116	5	10111010111110110010000010100010	2
117	5	11001101110111110000010001001100	2
118	5	00110110110000110011110010010011	2
119	5	00010101101010001110101001010111	2
120	5	00010011110010001110110000110111	1
121	5	10010110110000110011110010010110	1
122	5	10000000101010001110101011111110	1
123	5	01100000000001101001111111111001	1
124	5	01000110011000101011100110011101	1
125	5	00010111010111110000010100010111	1
126	5	00000001001101110001001101111111	1
127	5	11111001111101101001000001100000	1
128	5	00000111111000001111100000011111	1
129	5	11010100110111010100010011010100	1
130	5	11101011101111101000001000101000	1
131	5	11100001000111101000011101111000	1
132	5	00000110011000001111100110011111	1

1. Reconf function = a'b'c' + abc' + ab'c + a'bc
   f1 = b'c' + bc
   f2 = bc' + b'c
2. Reconf function = b'c' + a'c' + a'b'
   f1 = c' + b'
   f2 = b'c'
3. Reconf function = ab'c'd'e' + a'bc'd'e' + a'b'cd'e' + abcd'e' + a'b'c'de' + abc'de' + ab'cde' + a'bcde' + a'b'c'd'e + abc'd'e + ab'cd'e + a'bcd'e + ab'c'de + a'bc'de + a'b'cde + abcde
   f1 = bc'd'e' + b'cd'e' + b'c'de' + bcde' + b'c'd'e + bcd'e + bc'de + b'cde
   f2 = b'c'd'e' + bcd'e' + bc'de' + b'cde' + bc'd'e + b'cd'e + b'c'de + bcde
4. Reconf function = b'c + a'c + a'b'
   f1 = c + b'
   f2 = b'c
5. Reconf function = bc + a'c + a'b f1 = c + b
   f2 = bc
6. Reconf function = b'c + ac + ab'
   f1 = b'c
   f2 = c + b'

7. Reconf function = b’c’d’e’ + a’c’d’e’ + a’b’d’e’ + bcde’ + acde’ + abde’ + bcd’e + acd’e + abd’e + b’c’de + a’c’de + a’b’de
   f1 = c’d’e’ + b’d’e’ + bcde’ + bcd’e + c’de + b’de
   f2 = b’c’d’e’ + cde’ + bde’ + cd’e + bd’e + b’c’de
8. Reconf function = ac’d’e’ + ab’d’e’ + a’cde’ + a’bde’ + ab’c’e’ + a’bce’ + a’c’d’e + a’b’d’e + acde + abde + a’b’c’e + abce
   f1 = cde’ + bde’ + bce’ + c’d’e + b’d’e + b’c’e
   f2 = c’d’e’ + b’d’e’ + b’c’e’ + cde + bde + bce
9. Reconf function = bcd’e’ + a’cd’e’ + a’bd’e’ + b’c’de’ + ac’de’ + ab’de’ + b’c’d’e + ac’d’e + ab’d’e + bcde + a’cde + a’bde
   f1 = cd’e’ + bd’e’ + b’c’de’ + b’c’d’e + cde + bde
   f2 = bcd’e’ + c’de’ + b’de’ + c’d’e + b’d’e + bcde
10. Reconf function = bc’d’e’ + ac’d’e’ + abd’e’ + b’cde’ + a’cde’ + a’b’de’ + b’cd’e + a’cd’e + a’b’d’e + bc’de + ac’de + abde
    f1 = bc’d’e’ + cde’ + b’de’ + cd’e + b’d’e + bc’de
    f2 = c’d’e’ + bd’e’ + b’cde’ + b’cd’e + c’de + bde
11. Reconf function = bc’d’e’ + a’c’d’e’ + a’bd’e’ + b’cde’ + acde’ + ab’de’ + b’cd’e + acd’e + ab’d’e + bc’de + a’c’de + a’bde
    f1 = c’d’e’ + bd’e’ + b’cde’ + b’cd’e + c’de + bde
    f2 = bc’d’e’ + cde’ + b’de’ + cd’e + b’d’e + bc’de
12. Reconf function = a’b’d’ + abd’ + ab’c’ + a’bc’
    f1 = b’d’ + bc’
    f2 = bd’ + b’c’
13. Reconf function = d’e’ + b’c’e’ + a’c’e’ + a’b’e’ + b’c’d’ + a’c’d’ + a’b’d’
    f1 = d’e’ + c’e’ + b’e’ + c’d’ + b’d’
    f2 = d’e’ + b’c’e’ + b’c’d’
14. Reconf function = d’e + bce + ace + abe + bcd’ + acd’ + abd’
    f1 = d’e + bce + bcd’
    f2 = d’e + ce + be + cd’ + bd’
15. Reconf function = a’b’d’ + abd’ + ab’c + a’bc
    f1 = b’d’ + bc
    f2 = bd’ + b’c
16. Reconf function = a’cd’e’ + a’bd’e’ + ac’de’ + ab’de’ + ab’c’e’ + a’bce’ + acd’e + abd’e + a’c’de + a’b’de + a’b’c’e + abce
    f1 = cd’e’ + bd’e’ + bce’ + c’de + b’de + b’c’e
    f2 = c’de’ + b’de’ + b’c’e’ + cd’e + bd’e + bce
17. Reconf function = a’c’d + b’d + b’c’ + a’b’
    f1 = c’d + b’
    f2 = b’d + b’c’
18. Reconf function = bc’d’e’ + a’bd’e’ + b’cde’ + ab’de’ + a’bc’e’ + ab’ce’ + b’c’d’e + a’b’d’e + bcde + abde + a’b’c’e + abce
    f1 = bd’e’ + b’cde’ + bc’e’ + b’d’e + bcde + b’c’e
    f2 = bc’d’e’ + b’de’ + b’ce’ + b’c’d’e + bde + bce

19. Reconf function = acd'e' + ab'd'e' + a'c'de' + a'bde' + a'bc'e' + ab'ce' + a'cd'e + a'b'd'e + ac'de + abde + abc'e + a'b'ce
    f1 = c'de' + bde' + bc'e' + cd'e + b'd'e + b'ce
    f2 = cd'e' + b'd'e' + b'ce' + c'de + bde + bc'e
20. Reconf function = de + b'c'e + a'c'e + a'b'e + b'c'd + a'c'd + a'b'd
    f1 = de + c'e + b'e + c'd + b'd
    f2 = de + b'c'e + b'c'd
21. Reconf function = a'c'd'e' + a'bd'e' + acde' + ab'de' + a'bc'e' + ab'ce' + ac'd'e + abd'e + a'cde + a'b'de + abc'e + a'b'ce
    f1 = c'd'e' + bd'e' + bc'e' + cde + b'de + b'ce
    f2 = cde' + b'de' + b'ce' + c'd'e + bd'e + bc'e
22. Reconf function = d'e + a'b'c'e + abc'e + ab'ce + a'bce + a'b'c'd' + abc'd' + ab'cd' + a'bcd'
    f1 = d'e + b'c'e + bce + b'c'd' + bcd'
    f2 = d'e + bc'e + b'ce + bc'd' + b'cd'
23. Reconf function = b'cd'e' + ab'd'e' + bcde' + abde' + bc'd'e + a'bd'e + b'c'de + a'b'de + a'bc'd' + ab'cd' + a'b'c'd + abcd
    f1 = b'cd'e' + bcde' + bd'e + b'de + bc'd' + b'c'd
    f2 = b'd'e' + bde' + bc'd'e + b'c'de + b'cd' + bcd
24. Reconf function = c'd'e + a'd'e + b'c'e + a'b'e + a'c'd' + b'd' + a'b'c'
    f1 = d'e + b'e + c'd' + b'd' + b'c'
    f2 = c'd'e + b'c'e + b'd'
25. Reconf function = ab'c'd' + a'cd' + a'bc'd + acd + a'b'c + abc
    f1 = cd' + bc'd + b'c
    f2 = b'c'd' + cd + bc
26. Reconf function = ab'd' + a'bd' + a'b'c' + abc'
    f1 = bd' + b'c'
    f2 = b'd' + bc'
27. Reconf function = ab'd' + a'bd' + a'b'c + abc
    f1 = bd' + b'c
    f2 = b'd' + bc
28. Reconf function = cde + bde + bce + ae + acd + abd + abc
    f1 = cde + bde + bce
    f2 = e + cd + bd + bc
29. Reconf function = cde + a'de + b'ce + a'b'e + a'cd + b'd + a'b'c
    f1 = de + b'e + cd + b'd + b'c
    f2 = cde + b'ce + b'd
30. Reconf function = de + ab'c'e + a'bc'e + a'b'ce + abce + ab'c'd + a'bc'd + a'b'cd + abcd
    f1 = de + bc'e + b'ce + bc'd + b'cd
    f2 = de + b'c'e + bce + b'c'd + bcd
31. Reconf function = a'b'c'd' + abcd' + ac'd + a'cd + ab'd + a'bd
    f1 = b'c'd' + cd + bd
    f2 = bcd' + c'd + b'd

32. Reconf function = bc'd'e' + ab'de' + b'cd'e + a'bde + abc'd' + a'b'cd' + b'c'd + bcd
    f1 = bc'd'e' + bde + b'cd' + b'c'd + bcd
    f2 = b'de' + b'cd'e + bc'd' + b'c'd + bcd
33. Reconf function = bc'de' + b'cde' + abc'e' + ab'ce' + b'c'de + bcde + ab'c'e + abce + ad
    f1 = bc'de' + b'cde' + b'c'de + bcde
    f2 = bc'e' + b'ce' + b'c'e + bce + d
34. Reconf function = abc'd' + a'b'd' + ab'cd + a'bd + a'b'c' + a'bc
    f1 = b'd' + bd + b'c' + bc
    f2 = bc'd' + b'cd
35. Reconf function = b'd'e + a'c'e + c'd' + a'b'd' + b'c'
    f1 = c'e + c'd' + b'd' + b'c'
    f2 = b'd'e + c'd' + b'c'
36. Reconf function = acd + bd + bc + ab
    f1 = bd + bc
    f2 = cd + b
37. Reconf function = b'c'de' + bcde' + ab'c'e' + abce' + bc'de + b'cde + abc'e + ab'ce + ad
    f1 = b'c'de' + bcde' + bc'de + b'cde
    f2 = b'c'e' + bce' + bc'e + b'ce + d
38. Reconf function = ab'cd' + a'bd' + abc'd + a'b'd + a'b'c' + a'bc
    f1 = bd' + b'd + b'c' + bc
    f2 = b'cd' + bc'd
39. Reconf function = a'c'd' + acd' + a'b'd' + abd' + ab'c'd + a'bcd
    f1 = c'd' + b'd' + bcd
    f2 = cd' + bd' + b'c'd
40. Reconf function = a'b'c'd' + abd' + a'bcd + ab'd + abc' + ab'c
    f1 = b'c'd' + bcd
    f2 = bd' + b'd + bc' + b'c
41. Reconf function = de + b'ce + ace + ab'e + b'cd + acd + ab'd
    f1 = de + b'ce + b'cd
    f2 = de + ce + b'e + cd + b'd
42. Reconf function = de + a'b'c'e + abc'e + ab'ce + a'bce + a'b'c'd + abc'd + ab'cd + a'bcd
    f1 = de + b'c'e + bce + b'c'd + bcd
    f2 = de + bc'e + b'ce + bc'd + b'cd
43. Reconf function = d'e + b'c'e + a'c'e + a'b'e + b'c'd' + a'c'd' + a'b'd'
    f1 = d'e + c'e + b'e + c'd' + b'd'
    f2 = d'e + b'c'e + b'c'd'
44. Reconf function = d'e + ab'c'e + a'bc'e + a'b'ce + abce + ab'c'd' + a'bc'd' + a'b'cd' + abcd'
    f1 = d'e + bc'e + b'ce + bc'd' + b'cd'
    f2 = d'e + b'c'e + bce + b'c'd' + bcd'

45. Reconf function = b’cd’e’ + a’cd’e’ + bc’de’ + ac’de’ + abc’e’ + a’b’ce’ + b’c’d’e + a’c’d’e + bcde + acde + a’b’c’e + abce
    f1 = cd’e’ + bc’de’ + b’ce’ + c’d’e + bcde + b’c’e
    f2 = b’cd’e’ + c’de’ + bc’e’ + b’c’d’e + cde + bce
46. Reconf function = d’e + bc’e + ac’e + abe + bc’d’ + ac’d’ + abd’
    f1 = d’e + bc’e + bc’d’
    f2 = d’e + c’e + be + c’d’ + bd’
47. Reconf function = cd’ + bd’ + ad’ + abc
    f1 = cd’ + bd’
    f2 = d’ + bc
48. Reconf function = bcd + a’d + a’c + a’b
    f1 = d + c + b
    f2 = bcd
49. Reconf function = ab’d’e’ + a’bd’e’ + ab’c’e’ + a’bc’e’ + a’b’de + abde + a’b’ce + abce + ab’c’d’ + a’bc’d’ + a’b’cd + abcd
    f1 = bd’e’ + bc’e’ + b’de + b’ce + bc’d’ + b’cd
    f2 = b’d’e’ + b’c’e’ + bde + bce + b’c’d’ + bcd
50. Reconf function = a’b’d’e’ + abc’e’ + a’bde + ab’ce + abc’d’ + ab’cd + a’b’c’ + a’bc
    f1 = b’d’e’ + bde + b’c’ + bc
    f2 = bc’e’ + b’ce + bc’d’ + b’cd
51. Reconf function = b’c’d’e’ + ab’d’e’ + bcde’ + a’bde’ + ab’c’e’ + a’bce’ + bc’d’e + abd’e + b’cde + a’b’de + abc’e + a’b’ce
    f1 = b’c’d’e’ + bde’ + bce’ + bc’d’e + b’de + b’ce
    f2 = b’d’e’ + bcde’ + b’c’e’ + bd’e + b’cde + bc’e
52. Reconf function = de + b’ce + a’ce + a’b’e + b’cd + a’cd + a’b’d
    f1 = de + ce + b’e + cd + b’d
    f2 = de + b’ce + b’cd
53. Reconf function = c’d’ + a’bd’ + bc’ + a’c’
    f1 = bd’ + c’
    f2 = c’d’ + bc’
54. Reconf function = a’b’c’d’ + abd’ + ab’cd + a’bd + abc’ + a’bc
    f1 = b’c’d’ + bd + bc
    f2 = bd’ + b’cd + bc’
55. Reconf function = a’c’d’ + acd’ + ab’d’ + a’bd’ + abc’d + a’b’cd
    f1 = c’d’ + bd’ + b’cd
    f2 = cd’ + b’d’ + bc’d
56. Reconf function = d’e + b’ce + ace + ab’e + b’cd’ + acd’ + ab’d’
    f1 = d’e + b’ce + b’cd’
    f2 = d’e + ce + b’e + cd’ + b’d’
57. Reconf function = bcd + ad + ac + ab
    f1 = bcd
    f2 = d + c + b

58. Reconf function = a'cd' + b'd' + b'c + a'b'
    f1 = cd' + b'
    f2 = b'd' + b'c
59. Reconf function = de + b'c'e + ac'e + ab'e + b'c'd + ac'd + ab'd
    f1 = de + b'c'e + b'c'd
    f2 = de + c'e + b'e + c'd + b'd
60. Reconf function = de + bc'e + ac'e + abe + bc'd + ac'd + abd
    f1 = de + bc'e + bc'd
    f2 = de + c'e + be + c'd + bd
61. Reconf function = a'c'd'e' + acd'e' + a'b'c'e' + ab'ce' + ac'de + a'cde + abc'e + a'bce + a'b'c'd' + ab'cd' + abc'd + a'bcd
    f1 = c'd'e' + b'c'e' + cde + bce + b'c'd' + bcd
    f2 = cd'e' + b'ce' + c'de + bc'e + b'cd' + bc'd
62. Reconf function = b'cd'e' + ab'd'e' + bc'de' + a'bde' + a'bc'e' + ab'ce' + bcd'e + abd'e + b'c'de + a'b'de + a'b'c'e + abce
    f1 = b'cd'e' + bde' + bc'e' + bcd'e + b'de + b'c'e
    f2 = b'd'e' + bc'de' + b'ce' + bd'e + b'c'de + bce
63. Reconf function = ac'd' + a'cd' + a'b'd' + abd' + a'bc'd + ab'cd
    f1 = cd' + b'd' + bc'd
    f2 = c'd' + bd' + b'cd
64. Reconf function = acd + b'd + b'c + ab'
    f1 = b'd + b'c
    f2 = cd + b'
65. Reconf function = a'c'd' + ab'd' + ac'd + a'b'd
    f1 = c'd' + b'd
    f2 = b'd' + c'd
66. Reconf function = b'c'd'e' + bcd'e' + a'b'c'e' + a'bce' + bc'de + b'cde + abc'e + ab'ce + a'b'c'd' + a'bcd' + abc'd + ab'cd
    f1 = b'c'e' + bce' + bc'de + b'cde + b'c'd' + bcd'
    f2 = b'c'd'e' + bcd'e' + bc'e + b'ce + bc'd + b'cd
67. Reconf function = c'd + a'b'd + b'c' + a'c'
    f1 = b'd + c'
    f2 = c'd + b'c'
68. Reconf function = abcd' + a'b'd' + ab'c'd + a'bd + a'bc' + a'b'c
    f1 = b'd' + bd + bc' + b'c
    f2 = bcd' + b'c'd
69. Reconf function = b'c'd + a'd + a'c' + a'b'
    f1 = d + c' + b'
    f2 = b'c'd
70. Reconf function = de' + bc'e' + ac'e' + abe' + bc'd + ac'd + abd
    f1 = de' + bc'e' + bc'd
    f2 = de' + c'e' + be' + c'd + bd
71. Reconf function = c'de + bde + bc'e + ae + ac'd + abd + abc'
    f1 = c'de + bde + bc'e
    f2 = e + c'd + bd + bc'

72. Reconf function = bc'd'e' + b'cd'e' + b'c'de' + bcde' + ae' + abc'd' + ab'cd' + ab'c'd + abcd
    f1 = bc'd'e' + b'cd'e' + b'c'de' + bcde'
    f2 = e' + bc'd' + b'cd' + b'c'd + bcd
73. Reconf function = cd + abd + bc + ac
    f1 = cd + bc
    f2 = bd + c
74. Reconf function = c'd'e' + a'd'e' + a'c'e' + be' + bc'd' + a'bd' + a'bc'
    f1 = d'e' + c'e' + be' + bd' + bc'
    f2 = c'd'e' + be' + bc'd'
75. Reconf function = cde + bde + bce + a'e + a'cd + a'bd + a'bc
    f1 = e + cd + bd + bc
    f2 = cde + bde + bce
76. Reconf function = cd'e + bd'e + bce + ae + acd' + abd' + abc
    f1 = cd'e + bd'e + bce
    f2 = e + cd' + bd' + bc
77. Reconf function = b'cd' + a'd' + a'c + a'b'
    f1 = d' + c + b'
    f2 = b'cd'
78. Reconf function = c'd'e + bd'e + bc'e + ae + ac'd' + abd' + abc'
    f1 = c'd'e + bd'e + bc'e
    f2 = e + c'd' + bd' + bc'
79. Reconf function = ab'de' + a'bde' + ab'ce' + a'bce' + a'b'd'e + abd'e + a'b'c'e + abc'e + a'b'c'd' + abc'd' + ab'cd + a'bcd
    f1 = bde' + bce' + b'd'e + b'c'e + b'c'd' + bcd
    f2 = b'de' + b'ce' + bd'e + bc'e + bc'd' + b'cd
80. Reconf function = a'c'd' + acd' + ab'c' + a'b'c
    f1 = c'd' + b'c
    f2 = cd' + b'c'
81. Reconf function = de' + b'ce' + ace' + ab'e' + b'cd + acd + ab'd
    f1 = de' + b'ce' + b'cd
    f2 = de' + ce' + b'e' + cd + b'd
82. Reconf function = b'c'd'e + bcd'e + bc'de + b'cde + ae + ab'c'd' + abcd' + abc'd + ab'cd
    f1 = b'c'd'e + bcd'e + bc'de + b'cde
    f2 = e + b'c'd' + bcd' + bc'd + b'cd
83. Reconf function = ab'cd' + a'bd' + a'b'c'd + abd + abc' + a'bc
    f1 = bd' + b'c'd + bc
    f2 = b'cd' + bd + bc'
84. Reconf function = bcd'e' + a'bd'e' + b'c'de' + ab'de' + ab'c'e' + a'bce' + b'cd'e + a'b'd'e + bc'de + abde + abc'e + a'b'ce
    f1 = bd'e' + b'c'de' + bce' + b'd'e + bc'de + b'ce
    f2 = bcd'e' + b'de' + b'c'e' + b'cd'e + bde + bc'e
85. Reconf function = c'd'e + b'd'e + b'c'e + ae + ac'd' + ab'd' + ab'c'
    f1 = c'd'e + b'd'e + b'c'e
    f2 = e + c'd' + b'd' + b'c'

86. Reconf function = de + bc'e + a'c'e + a'be + bc'd + a'c'd + a'bd
    f1 = de + c'e + be + c'd + bd
    f2 = de + bc'e + bc'd
87. Reconf function = cde' + bde' + bce' + ae' + acd + abd + abc
    f1 = cde' + bde' + bce'
    f2 = e' + cd + bd + bc
88. Reconf function = b'cd'e' + acd'e' + b'c'de' + ac'de' + b'c'd'e + ac'd'e + b'cde + acde + ab'
    f1 = b'cd'e' + b'c'de' + b'c'd'e + b'cde
    f2 = cd'e' + c'de' + c'd'e + cde + b'
89. Reconf function = de + bce + a'ce + a'be + bcd + a'cd + a'bd
    f1 = de + ce + be + cd + bd
    f2 = de + bce + bcd
90. Reconf function = c'de + b'de + b'c'e + ae + ac'd + ab'd + ab'c'
    f1 = c'de + b'de + b'c'e
    f2 = e + c'd + b'd + b'c'
91. Reconf function = bc'd'e' + a'bd'e' + b'c'de' + a'b'de' + b'cd'e + ab'd'e + bcde + abde + a'bc'd' + ab'cd' + a'b'c'd + abcd
    f1 = bd'e' + b'de' + b'cd'e + bcde + bc'd' + b'c'd
    f2 = bc'd'e' + b'c'de' + b'd'e + bde + b'cd' + bcd
92. Reconf function = ac'd' + a'bcd' + a'c'd + ab'cd + a'b'c' + abc'
    f1 = bcd' + c'd + b'c'
    f2 = c'd' + b'cd + bc'
93. Reconf function = a'b'cd' + abd' + ab'c'd + a'bd + a'bc' + abc
    f1 = b'cd' + bd + bc'
    f2 = bd' + b'c'd + bc
94. Reconf function = c'd + abd + bc' + ac'
    f1 = c'd + bc'
    f2 = bd + c'
95. Reconf function = de' + b'c'e' + ac'e' + ab'e' + b'c'd + ac'd + ab'd
    f1 = de' + b'c'e' + b'c'd
    f2 = de' + c'e' + b'e' + c'd + b'd
96. Reconf function = d'e + b'c'e + ac'e + ab'e + b'c'd' + ac'd' + ab'd'
    f1 = d'e + b'c'e + b'c'd'
    f2 = d'e + c'e + b'e + c'd' + b'd'
97. Reconf function = ac'd' + a'b'cd' + a'c'd + abcd + ab'c' + a'bc'
    f1 = b'cd' + c'd + bc'
    f2 = c'd' + bcd + b'c'
98. Reconf function = a'cd + b'd + b'c + a'b'
    f1 = cd + b'
    f2 = b'd + b'c
99. Reconf function = bcd' + a'd' + a'c + a'b
    f1 = d' + c + b
    f2 = bcd'

100. Reconf function = ac'd + b'd + b'c' + ab'
    f1 = b'd + b'c'
    f2 = c'd + b'
101. Reconf function = cde' + a'de' + b'ce' + a'b'e' + a'cd + b'd + a'b'c
    f1 = de' + b'e' + cd + b'd + b'c
    f2 = cde' + b'ce' + b'd
102. Reconf function = cd'e' + c'de' + ab'e' + a'be' + a'b'c'd'e + abc'd'e + a'b'cde + abcde
    f1 = cd'e' + c'de' + be' + b'c'd'e + b'cde
    f2 = cd'e' + c'de' + b'e' + bc'd'e + bcde
103. Reconf function = bde + a'de + bc'e + a'c'e + bc'd + a'c'd + a'b
    f1 = de + c'e + c'd + b
    f2 = bde + bc'e + bc'd
104. Reconf function = c'd + a'bd + bc' + a'c'
    f1 = bd + c'
    f2 = c'd + bc'
105. Reconf function = c'd + ab'd + b'c' + ac'
    f1 = c'd + b'c'
    f2 = b'd + c'
106. Reconf function = b'cd' + ad' + ac + ab'
    f1 = b'cd'
    f2 = d' + c + b'
107. Reconf function = c'd + b'd + ad + ab'c'
    f1 = c'd + b'd
    f2 = d + b'c'
108. Reconf function = ac'd'e' + ab'd'e' + a'c'de' + a'b'de' + a'cd'e + a'bd'e + acde + abde + ab'c'd' + a'bcd' + a'b'c'd + abcd
    f1 = c'de' + b'de' + cd'e + bd'e + bcd' + b'c'd
    f2 = c'd'e' + b'd'e' + cde + bde + b'c'd' + bcd
109. Reconf function = a'c'd + bd + bc' + a'b
    f1 = c'd + b
    f2 = bd + bc'
110. Reconf function = b'd'e + a'd'e + ce + a'b'e + b'cd' + a'cd' + a'b'c
    f1 = d'e + ce + b'e + cd' + b'c
    f2 = b'd'e + ce + b'cd'
111. Reconf function = b'c'd + bcd + a'bc' + a'b'c
    f1 = d + bc' + b'c
    f2 = b'c'd + bcd
112. Reconf function = a'cd + bd + bc + a'b
    f1 = cd + b
    f2 = bd + bc
113. Reconf function = cd + a'b'd + b'c + a'c
    f1 = b'd + c
    f2 = cd + b'c

114. Reconf function = cd'e + bd'e + bce + a'e + a'cd' + a'bd' + a'bc
 f1 = e + cd' + bd' + bc
 f2 = cd'e + bd'e + bce
115. Reconf function = a'b'd'e' + abde' + a'c'e' + ace' + ac'd'e + a'cde + ab'c'e + a'bce
 f1 = b'd'e' + c'e' + cde + bce
 f2 = bde' + ce' + c'd'e + b'c'e
116. Reconf function = cd'e + b'd'e + b'ce + ae + acd' + ab'd' + ab'c
 f1 = cd'e + b'd'e + b'ce
 f2 = e + cd' + b'd' + b'c
117. Reconf function = c'd'e + a'd'e + a'c'e + be + bc'd' + a'bd' + a'bc'
 f1 = d'e + c'e + be + bd' + bc'
 f2 = c'd'e + be + bc'd'
118. Reconf function = bc'de' + a'b'ce' + bcd'e + ab'c'e + b'c'd' + abcd' + a'bc'd + b'cd
 f1 = b'ce' + bcd'e + b'c'd' + bc'd + b'cd
 f2 = bc'de' + b'c'e + b'c'd' + bcd' + b'cd
119. Reconf function = a'd'e' + ade' + ab'c'e' + a'bce' + acd'e + abd'e + a'c'de + a'b'de
 f1 = d'e' + bce' + c'de + b'de
 f2 = de' + b'c'e' + cd'e + bd'e
120. Reconf function = b'd'e' + bde' + a'bc'e' + ab'ce' + bcd'e + abd'e + b'c'de + a'b'de
 f1 = b'd'e' + bde' + bc'e' + bcd'e + b'de
 f2 = b'd'e' + bde' + b'ce' + bd'e + b'c'de
121. Reconf function = bc'de' + b'cde' + a'bc'e' + a'b'ce' + b'c'd'e + bcd'e + ab'c'e + abce + ab'c'd' + abcd' + a'bc'd + a'b'cd
 f1 = bc'e' + b'ce' + b'c'd'e + bcd'e + bc'd + b'cd
 f2 = bc'de' + b'cde' + b'c'e + bce + b'c'd' + bcd'
122. Reconf function = cd'e' + bd'e' + bce' + ae' + acd' + abd' + abc
 f1 = cd'e' + bd'e' + bce'
 f2 = e' + cd' + bd' + bc
123. Reconf function = cd'e' + c'de' + a'b'e' + abe' + ab'c'd'e + a'bc'd'e + ab'cde + a'bcde
 f1 = cd'e' + c'de' + b'e' + bc'd'e + bcde
 f2 = cd'e' + c'de' + be' + b'c'd'e + b'cde
124. Reconf function = a'c'd'e' + acde' + a'b'e' + abe' + ab'd'e + a'bde + ab'c'e + a'bce
 f1 = c'd'e' + b'e' + bde + bce
 f2 = cde' + be' + b'd'e + b'c'e
125. Reconf function = c'd'e + a'd'e + b'c'e + a'b'e + b'c'd' + a'b'd' + a'c'
 f1 = d'e + b'e + b'd' + c'
 f2 = c'd'e + b'c'e + b'c'd'

126. Reconf function = c'd'e' + a'd'e' + b'c'e' + a'b'e' + a'c'd' + b'd' + a'b'c'
f1 = d'e' + b'e' + c'd' + b'd' + b'c'
f2 = c'd'e' + b'c'e' + b'd'
127. Reconf function = ab'd'e + a'bd'e + a'b'de + abde + ce + ab'cd' + a'bcd' + a'b'cd + abcd
f1 = bd'e + b'de + ce + bcd' + b'cd
f2 = b'd'e + bde + ce + b'cd' + bcd
128. Reconf function = c'd'e' + cde' + abc'e' + a'b'ce' + bcd'e + acd'e + b'c'de + a'c'de
f1 = c'd'e' + cde' + b'ce' + bcd'e + c'de
f2 = c'd'e' + cde' + bc'e' + cd'e + b'c'de
129. Reconf function = bd'e + a'd'e + bce + a'ce + bcd' + a'cd' + a'b
f1 = d'e + ce + cd' + b
f2 = bd'e + bce + bcd'
130. Reconf function = bc'd'e + b'cd'e + b'c'de + bcde + ae + abc'd' + ab'cd' + ab'c'd + abcd
f1 = bc'd'e + b'cd'e + b'c'de + bcde
f2 = e + bc'd' + b'cd' + b'c'd + bcd
131. Reconf function = b'cd'e' + a'cd'e' + b'c'de' + a'c'de' + bc'd'e + ac'd'e + bcde + acde + abc'd' + a'b'cd' + a'b'c'd + abcd
f1 = cd'e' + c'de' + bc'd'e + bcde + b'cd' + b'c'd
f2 = b'cd'e' + b'c'de' + c'd'e + cde + bc'd' + bcd
132. Reconf function = c'd'e' + cde' + a'b'e' + abe' + ab'cd'e + a'bcd'e + ab'c'de + a'bc'de
f1 = c'd'e' + cde' + b'e' + bcd'e + bc'de
f2 = c'd'e' + cde' + be' + b'cd'e + b'c'de

Appendix C
Layout Extraction File (`.lef`) for SiNW RFETs

```
VERSION 5.7;
BUSBITCHARS "[]";
DIVIDERCHAR "/";

# Metal layer Description.
LAYER metal1
TYPE ROUTING;
DIRECTION HORIZONTAL;
PITCH 0.104 ;
OFFSET 0.052 ;
WIDTH 0.040;
SPACING 0.040;
RESISTANCE RPERSQ 0.07;
CAPACITANCE CPERSQDIST 3e-05;
END metal1

LAYER via1
TYPE CUT;
SPACING 0.040;
END via1

VIARULE via1_array GENERATE
LAYER metal1;
 2*ENCLOSURE 0.01 = metal1 rect
ENCLOSURE 0.01 0.01;
LAYER metal2;
ENCLOSURE 0.01 0.01;
LAYER via1;
#   bottom left point, top right point
```


S. Rai, A. Kumar, *Design Automation and Applications for Emerging Reconfigurable Nanotechnologies*, https://doi.org/10.1007/978-3-031-37924-6

```
RECT -0.0100 -0.0100 0.0100 0.0100;
#   center-to-center spacing, X BY Y
SPACING 0.04 BY 0.04;
END via1_array

MACRO inv1
CLASS CORE;
ORIGIN 0 0;
FOREIGN inv1 0 0;
SIZE 0.416 BY 0.72;
SYMMETRY X Y;
SITE 104cpp_9t;
PIN VSS
DIRECTION INOUT;
USE GROUND;
SHAPE ABUTMENT;
PORT
LAYER metal1;
RECT 0 -0.02 0.416 0.02;
RECT 0.24 -0.02 0.3 0.125;
END
END VSS
PIN VDD
DIRECTION INOUT;
USE POWER;
SHAPE ABUTMENT;
PORT
LAYER metal1;
RECT 0 0.7 0.416 0.74;
RECT 0.116 0.595 0.176 0.74;
END
END VDD
PIN a
DIRECTION INPUT;
USE SIGNAL;
ANTENNAMODEL OXIDE1;
ANTENNAGATEAREA 0.008 LAYER metal1;
PORT
LAYER metal1;
RECT 0.138 0.42 0.322 0.46;
RECT 0.138 0.23 0.198 0.46;
END
END a
PIN O
DIRECTION OUTPUT;
```

```
USE SIGNAL;
ANTENNADIFFAREA 0.0248 LAYER metal1;
PORT
LAYER metal1;
RECT 0.344 0.5 0.384 0.644;
RECT 0.032 0.5 0.384 0.54;
RECT 0.032 0.14 0.072 0.54;
END
END O
END inv1

END LIBRARY
```

Appendix D
Liberty (.lib) File for SiNW RFETs

```
library(hpsnlib_rfet_tc_1d80V_25C) {
delay_model : table_lookup;
library_features(report_delay_calculation, report_power_calculation);
time_unit : 1ns;
voltage_unit : 1V;
current_unit : 1uA;
capacitive_load_unit(1, ff);
pulling_resistance_unit : 1kohm;
leakage_power_unit : 1pW;
input_threshold_pct_fall : 50;
input_threshold_pct_rise : 50;
output_threshold_pct_fall : 50;
output_threshold_pct_rise : 50;
slew_derate_from_library : 0.5;
slew_lower_threshold_pct_fall : 30;
slew_lower_threshold_pct_rise : 30;
slew_upper_threshold_pct_fall : 70;
slew_upper_threshold_pct_rise : 70;
nom_process : 1;
nom_temperature : 25;
nom_voltage : 1.8;
default_cell_leakage_power : 0;
default_fanout_load : 0;
default_inout_pin_cap : 0;
default_input_pin_cap : 0;
default_leakage_power_density : 0;
default_output_pin_cap : 0;
voltage_map(VDD, 1.8);
voltage_map(VSS, 0);

operating_conditions(tc_1d80V_25C) {
process : 1;
```


S. Rai, A. Kumar, *Design Automation and Applications for Emerging Reconfigurable Nanotechnologies*, https://doi.org/10.1007/978-3-031-37924-6

```
temperature : 25;
voltage : 1.8;
}
default_operating_conditions : tc_1d80V_25C;

input_voltage(default) {
vil : 0;
vih : 1.8;
vimin : 0;
vimax : 1.8;
}

output_voltage(default) {
vol : 0;
voh : 1.8;
vomin : 0;
vomax : 1.8;
}

sensitization(sensitization_2pins) {
pin_names(pin_0, pin_1);
vector(0, "0 0");
vector(1, "0 1");
vector(2, "1 0");
vector(3, "1 1");
}

sensitization(sensitization_3pins) {
pin_names(pin_0, pin_1, pin_2);
vector(0, "0 0 0");
vector(1, "0 0 1");
vector(2, "0 1 0");
vector(3, "0 1 1");
vector(4, "1 0 0");
vector(5, "1 0 1");
vector(6, "1 1 0");
}

sensitization(sensitization_4pins) {
pin_names(pin_0, pin_1, pin_2, pin_3);
vector(1, "0 0 0 1");
vector(3, "0 0 1 1");
vector(5, "0 1 0 1");
vector(6, "0 1 1 0");
vector(8, "1 0 0 0");
vector(10, "1 0 1 0");
vector(11, "1 0 1 1");
}
```

```
lu_table_template(ndw_ntin_nvolt_6x13) {
variable_1 : input_net_transition;
variable_2 : normalized_voltage;
index_1("1, 2, 3, 4, 5, 6");
index_2("1, 2, 3, 4, 5, 6, 7, 8, 9, 10, 11, 12, 13");
}

lu_table_template(tmg_ntin_oload_6x6) {
variable_1 : input_net_transition;
variable_2 : total_output_net_capacitance;
index_1("1, 2, 3, 4, 5, 6");
index_2("1, 2, 3, 4, 5, 6");
}

power_lut_template(pwr_tin_6) {
variable_1 : input_transition_time;
index_1("1, 2, 3, 4, 5, 6");
}

power_lut_template(pwr_tin_oload_6x6) {
variable_1 : input_transition_time;
variable_2 : total_output_net_capacitance;
index_1("1, 2, 3, 4, 5, 6");
index_2("1, 2, 3, 4, 5, 6");
}


}


cell(inv1) {
sensitization_master : sensitization_2pins;
pin_name_map(a, O);
area : 0.29664;
cell_leakage_power : 39.0394395;

leakage_power() {
related_pg_pin : "VDD";
value : "39.0394395";
}

pg_pin(VDD) {
voltage_name : VDD;
pg_type : primary_power;
}

pg_pin(VSS) {
voltage_name : VSS;
pg_type : primary_ground;
```

```
}

pin(a) {
capacitance : 0.399375;
direction : input;
driver_waveform_rise : driver_waveform_default_rise;
driver_waveform_fall : driver_waveform_default_fall;
fall_capacitance : 0.399323;
fall_capacitance_range(0.394542, 0.403334);
input_voltage : default;
max_transition : 36.0;
related_ground_pin : VSS;
related_power_pin : VDD;
rise_capacitance : 0.399427;
rise_capacitance_range(0.394542, 0.403334);
}

pin(O) {
direction : output;
function : "!a";
max_capacitance : 24.3;
max_transition : 28.8;
output_voltage : default;
related_ground_pin : VSS;
related_power_pin : VDD;
power_down_function : "!VDD+VSS";

internal_power() {
related_pg_pin : "VDD";
related_pin : "a";


rise_power(pwr_tin_oload_6x6) {
index_1("0.1, 0.4, 1.2, 4, 12, 36");
index_2("0.1, 0.4, 1, 3, 9, 27");
values("1.2986, 1.29712, 1.29539, 1.29339, 1.29835, 1.25778",\
"1.31769, 1.30484, 1.303, 1.29395, 1.28541, 1.27292",\
"1.30702, 1.32168, 1.36207, 1.33391, 1.29818, 1.24022",\
"1.29422, 1.29695, 1.30186, 1.38523, 1.41725, 1.27534",\
"1.37127, 1.30538, 1.3196, 1.3258, 1.40479, 1.52572",\
"1.47365, 1.46357, 1.4322, 1.44359, 1.39741, 2.22454");
}
}

timing() {
related_pin : "a";
timing_sense : negative_unate;
timing_type : combinational;
```

```
cell_fall(tmg_ntin_oload_6x6) {
index_1("0.1, 0.4, 1.2, 4, 12, 36");
index_2("0.1, 0.4, 1, 3, 9, 27");
values("0.167195, 0.243104, 0.382518, 0.872025, 2.32162, 6.58803",\
"0.308322, 0.380426, 0.522397, 1.02382, 2.48043, 6.77239",\
"0.456128, 0.643228, 0.82919, 1.32162, 2.80596, 7.10223",\
"1.04392, 1.36103, 1.78699, 2.61598, 4.09887, 8.4496",\
"1.42532, 2.15076, 3.13964, 4.8514, 7.52164, 12.1209",\
"4.06991, 4.78814, 6.38181, 8.12942, 14.8572, 22.7991");
}
wave_fall(1, 2);

cell_rise(tmg_ntin_oload_6x6) {
index_1("0.1, 0.4, 1.2, 4, 12, 36");
index_2("0.1, 0.4, 1, 3, 9, 27");
values("0.151786, 0.221898, 0.360403, 0.839933, 2.25578, 6.34873",\
"0.256233, 0.328478, 0.467311, 0.942082, 2.34323, 6.47274",\
"0.398507, 0.51469, 0.711545, 1.16766, 2.55962, 6.70035",\
"0.602373, 0.87441, 1.22842, 2.03108, 3.62982, 7.72331",\
"1.38653, 0.554688, 1.70894, 3.31935, 5.8124, 10.4466",\
"1.60079, 2.22685, 3.50898, 6.54292, 10.6509, 17.7695");
}
wave_rise(2, 1);

fall_transition(tmg_ntin_oload_6x6) {
index_1("0.1, 0.4, 1.2, 4, 12, 36");
index_2("0.1, 0.4, 1, 3, 9, 27");
values("0.255504, 0.37472, 0.680413, 1.64343, 4.43972, 12.5613",\
"0.269282, 0.385135, 0.650846, 1.57275, 4.50571, 12.8883",\
"0.871151, 0.893921, 0.816229, 1.62398, 4.36444, 12.9361",\
"1.21568, 1.56826, 1.74721, 2.26646, 4.44653, 12.7098",\
"4.20328, 3.37664, 3.61834, 4.59383, 6.10891, 12.7317",\
"4.24503, 5.22933, 5.98887, 10.1779, 11.9861, 18.1581");
}

rise_transition(tmg_ntin_oload_6x6) {
index_1("0.1, 0.4, 1.2, 4, 12, 36");
index_2("0.1, 0.4, 1, 3, 9, 27");
values("0.239664, 0.357893, 0.6526, 1.57325, 4.28203, 12.1578",\
"0.277735, 0.376114, 0.688798, 1.52355, 4.33037, 12.3678",\
"0.803509, 0.910699, 0.951496, 1.57349, 4.17542, 12.3875",\
"1.49813, 1.37551, 1.63522, 2.23217, 4.38152, 12.3751",\
"2.07502, 4.59827, 4.11215, 4.1386, 6.29992, 12.7308",\
"6.08436, 6.09375, 6.96149, 6.48099, 9.88257, 17.5781");
}
}
}
}
```

Index

S. Rai, A. Kumar, *Design Automation and Applications for Emerging Reconfigurable Nanotechnologies*, https://doi.org/10.1007/978-3-031-37924-6

Printed by Printforce, the Netherlands